THE FIRST RTs

THE FIRST RTs

A HISTORY OF RTs 1-151
BY ALAN TOWNSIN & TONY BEARD

Capital Transport

First published 1996

ISBN 185414 177 5

Published by Capital Transport Publishing, 38 Long Elmes, Harrow Weald, Middlesex

Printed by Staples Printers Rochester Ltd, Medway City Estate, Rochester, Kent

The front cover painting is by Barry Pearce from a photo by D.W.K. Jones

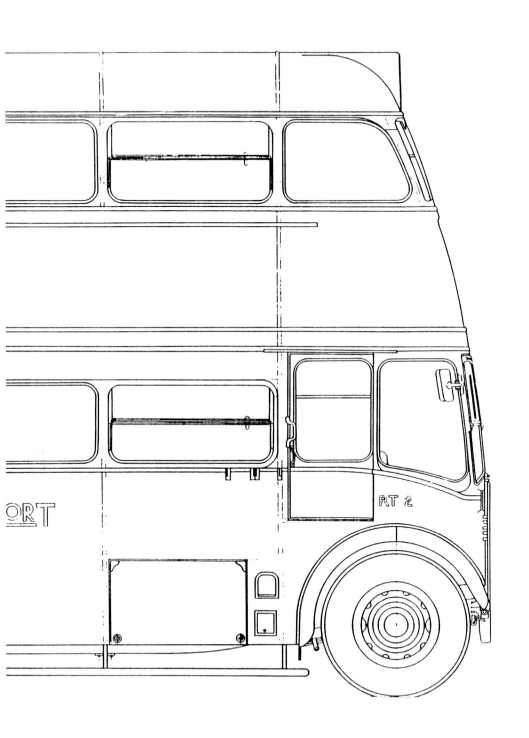
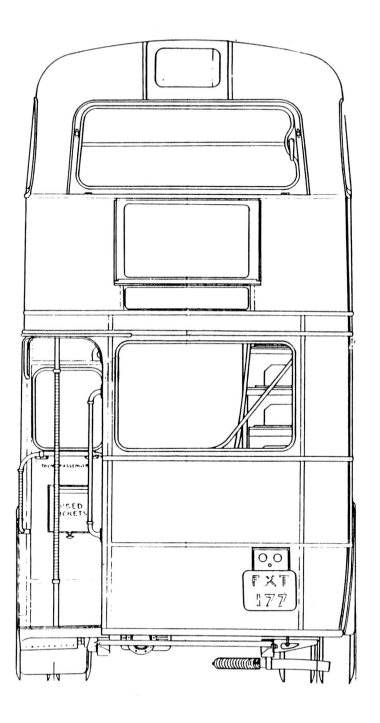

Preface

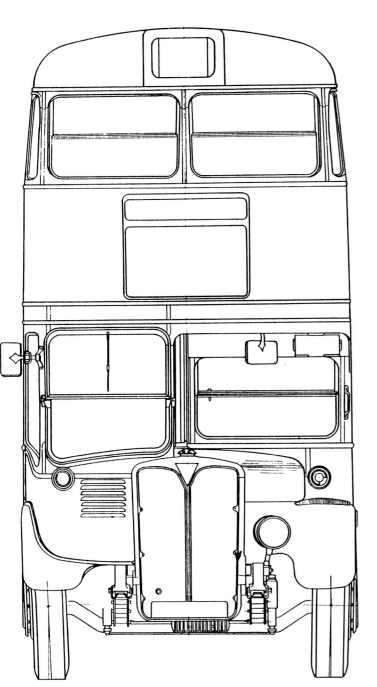

My interest in the London RT began when, at the age of thirteen, I saw a photograph in the Daily Telegraph of 14th July 1939 of the newly-built RT 1. My subsequent career started with an engineering slant, including joining AEC's drawing office staff in 1951, when the post-war version of the RT was still in production, and it would be impossible to thank all the people there who contributed to the store of information I gradually built up.

When specific work on this book began, it was fascinating how many new questions arose as study went deeper, and it was at this point that the 2RT2 Bus Preservation Group soon proved to be invaluable. This Group was formed in 1962 with the objective of purchasing and preserving one of the 'First RTs', and RT 113 was duly acquired from London Transport the following year. In addition to a thorough restoration programme, one of the Group's first members, the late Francis West, set about compiling complete life histories of each of the 151 vehicles from the official records then held at Chiswick Works. This material, augmented by painstaking further research by Tony Beard, Hon Secretary of the Group, has been of immense value as a detailed source of information.

Tony has proved to be an ideal colleague in the compilation of this book, both in the study of the subject and in stimulating fresh lines of thought in getting to the bottom of many interesting questions – there have been several occasions when the true story on subjects hitherto shrouded in mystery has emerged from his thorough analysis of the information that was to be found on the record cards. This tabular information, corrected as official records often require, will be published in full in the 2RT2 Handbook next year.

Prominent among those who have helped my studies of AEC and London bus matters over the years is Gavin Martin. He was also at AEC when I was there, though in the experimental department, after an earlier spell with London Transport, and his enquiring mind has always acted as a stimulus to getting to the root of many matters, such as the genealogy of the engine design adopted for the RT. Valuable information was obtained from the British Commercial Vehicle Trust Archive at Chorley, in which the AEC material is the speciality of Gordon Baron, and I must thank Brian Thackray, another AEC expert, for his assistance in this.

In regard to the London Transport side of the story, thanks are due to Ken Glazier and Colin Curtis, with emphasis on the 55 Broadway and Chiswick aspects respectively. John Gillham's records on RT 19's activities in 1940-42, made when he was at Chiswick, proved most useful and Bob Burrell supplied recollections from working there in the latter part of the war. From the garage perspective, Alan Pearce provided personal experiences of working on the 2RT2s as a bus mechanic at the two Putney garages in the post-war years and the early 1950s. Les Stitson assisted with route allocation details and Leon Daniels supplied material on the restoration of RT 1 after its purchase by Prince Marshall in 1978. Other valuable help has been provided by Lawrie Bowles, Alan Cross, Andrew Gilks, Gerald Mead and Alan Nightingale.

The authors and publisher hope that the end result does justice to the first examples of one of the top classic bus designs of all time.

Steventon, Hants, May 1996   Alan Townsin

Introduction

The formation of the London Passenger Transport Board in 1933 was a watershed in the story of public transport in London. It marked a departure from its provision by multiple concerns, even though they had been dominated by the Underground Group, a large private-enterprise company combine. The LPTB was a new type of organisation, set up by Act of Parliament as a free-standing public body deliberately given monopoly powers within its specified operating area, but charged with the provision of a comprehensive and efficient service. The quality of its rolling-stock was bound to come under close scrutiny as an indication of its commitment to improved standards.

Yet the new LPTB inherited much of its character from the Underground Group. In regard to buses, it was the London General Omnibus Co Ltd, the principal operator since horse-bus days but an Underground subsidiary since 1912, whose methods and ideas set the Board's initial standards.

The Associated Equipment Co Ltd, better known as AEC, had been set up as a separate subsidiary to take over the LGOC's bus chassis manufacturing activity, again in 1912, soon after the Underground took over. When the LPTB was formed, AEC became an independent business, even though a 10-year contract to supply the bulk of LPTB's bus chassis needs ensured continuity of involvement.

When the LPTB was formed, most people at all levels in the Underground Group continued in equivalent jobs. This applied right at the top, where Lord Ashfield continued as Chairman. The philosophy adopted by Ashfield and Frank Pick, his lieutenant for many years, was one of success based on high-quality public service, another element carried over to the new London Transport organisation and, if anything, magnified by the transfer.

In regard to bus engineering within the Central Area of the LPTB's territory, A.A.M. (Bill) Durrant was appointed Engineer (Central Buses). Earlier in 1933, he had become Chief Engineer of the LGOC. Bill Durrant had joined the LGOC in 1919, rising to Assistant Superintendent of Rolling Stock in 1929. After 1935, when his responsibility was extended to include the Country Area, his title became Chief Engineer (Buses and Coaches).

Lord Ashfield had been Chairman of Underground subsidiaries, including LGOC and AEC, again seeking high standards and taking an interest which extended right down to such details as the triangular radiator badge, which he had suggested. On the vehicle design side, Ashfield had watched the progress of a talented engineer, George John Rackham, himself a former LGOC man and one of the team responsible for the design of the B-type bus of 1910 but whose career after the 1914-18 war had taken him to America.

In 1922-26 Rackham had been Chief Engineer of the Yellow Coach Manufacturing Company, later to become the bus-building division of the giant General Motors combine. He came back to England to become Chief Engineer of Leyland Motors Ltd, promptly setting about the design of a new double-decker which was to make almost all other comparable models then on the market out of date when introduced as the Titan in October 1927.

His appointment as Chief Engineer of AEC was announced in July 1928 and, equally quickly, he set about designing a new range of chassis, beginning with a six-cylinder overhead-camshaft engine which, not surprisingly, shared many features with that of the Titan. The new chassis soon followed, and the first examples of the Regent double-decker, the Regal single-decker (both of which had two axles) and the three-axle Renown, were built in 1929. These were soon adopted by the LGOC as its types ST, T and LT respectively.

From 1929 to 1932, the LGOC and closely related companies placed in service 2,624 buses of these types (891 ST, 1,424 LT and 309 T). Most of them had the standard petrol engine in one or other of the alternative 6.1 or 7.4 litre sizes. However, quite extensive experimental work on oil (diesel) engines and an improved transmission system had been carried out. In October 1930, AEC became the first British maker to offer chassis with its own make of oil engine, but much further work needed to be done before a reliable engine was evolved. This was the result of co-operation with the Ricardo concern, the first resulting engine fitted in LT 643 in September 1931 being a landmark in this process. The production 8.8-litre A165 engine was soon giving good results in many of the later LT-class buses fitted with it from new in 1932.

The legal limit on gross weight of two-axle double-deckers was particularly tight at that date. That on three-axle (six-wheel) models was much less so, which was largely why most of the LGOC's and AEC's early work on oil engines, heavier than petrol units of comparable power, was centred on LT-class buses. Similar remarks applied to a lesser degree to the fluid flywheel and Wilson preselective gearbox, also tried mainly in some LT-class vehicles after initial tests on three Daimler CH6 buses bought for this purpose. By 1932, the latest standard LT-class double-decker seated 60.

However, the concept of a bus with similar capacity but on two axles was made possible by a change in maximum legal length from 25ft to 26ft for such buses which came into effect on 1st January 1932. AEC had shown that this could be done on an exhibit on a new 16ft 3in wheelbase version of the Regent chassis (in place of the previous 15ft 6½in) in anticipation of the change, at the Commercial Motor Show in November 1931.

It had obvious appeal and the LGOC produced a suitable body design, the result being the original STL, the initial batch beginning to enter service in January 1933. The 60-seat body design was of similar, almost vertical-fronted appearance to that on the Bluebird LT. These vehicles had petrol engines, partly because of the weight problem, but also because AEC's existing A165 engine necessitated the radiator being placed about 4½in further forward than standard because of its extra length. This would have required the body to be shortened by a similar amount to keep within the overall length limit, not impossible but unacceptable to LGOC on a maximum-capacity bus. On the LT-class buses this had not been a problem because they were well within the maximum of 30ft applicable with three axles.

Subsequently, the STL seating capacity was reduced to 56, and a sloping-fronted profile adopted for a batch of buses, STL 203-252, which was in hand when a meeting of the newly formed Board was held on 6th June 1933, three weeks before it took over responsibility on 1st July. Durrant was present and made a recommendation that the compression-ignition engine (another term for what is nowadays called the diesel) be adopted for the next batch of 100 double-deckers. In principle

this was agreed, but in practice all but the last eleven of these new STL buses went into service with existing petrol engines removed from LT-class double-deckers which were fitted with new 8.8-litre A165 oil engines. The eleven (STL 342-352) received pre-production examples of a new more compact oil engine of 7.7-litre nominal capacity, type A171.

In November 1934, a major turning point came with the appearance of STL 609, this being the first example of a variant with the A171 engine, AEC-built preselective gearbox (type D132) and a revised version of 56-seat body having a gently curving profile, setting new standards among British double-deckers. Over 1,900 generally similar buses were placed in service over the period up to mid-1938, taking the fleet numbers to STL 2515. Within this number there were further variations, mainly in regard to bodywork, and some improvements and experiments were carried out, but the basic mechanical design and outline came to be understood as the typical STL and the Board's first new standard type of bus.

The standard STLs were to prove generally very satisfactory buses, popular with drivers because of their lively performance and the ease of driving given by the preselective gearbox, even if the pedal action was rather heavy.

Yet there was a need for further improvements, both in search for greater efficiency and also to offset the growing pressure on drivers caused by increasing traffic congestion. Not to be under-rated was the expectation, already strong in LGOC days, but magnified by the formation of the LPTB, that London's standards in the design of its buses would be second to none.

This book tells the story of how these various needs were addressed in the design of the bus that was to become known as the RT, and how the prototype and first production batch paved the way for one of London's most famous bus types.

CHAPTER ONE

Planning to Prototype

There was a remarkable degree of unanimity among the major British makers and operators about the general specification of a double-deck bus in 1936-37. One clear reason for this was the restrictive nature of the legal limits on dimensions – there was universal agreement that the 26ft length and 7ft 6in width applicable to two-axle models was not only the maximum permissible but the minimum acceptable. Weight was not quite the problem it had been a little earlier, though the limit of $10\frac{1}{2}$ tons gross required close attention to design if 56 seated passengers were to be carried – there were many instances of buses which entered service with fewer seats because of this.

The three-axle double-deck motor bus was all but dead – versions with about 60 seats, as placed in service by the LGOC in 1932, showed too small a gain in seating capacity to justify their higher first cost and running expenses compared to a two-axle 56-seater. Attempting to go much beyond that in seating capacity again tended to run into weight problems, though the three-axle 70-seat trolleybus had become a well-established type, not least in London, where the fleet of such vehicles purchased for tram replacement was growing rapidly. The trolleybus weight allowance was not so restrictive.

The general layout of the 26ft double-deck motor bus was also remarkably standardised. The positioning of the engine longitudinally at the front, where it fitted conveniently in the relatively narrow space dictated by the need to allow for the steering movements of the front wheels, was virtually universal on motor vehicles of all sizes at that date, as was the transmission of drive to the rear wheels. AEC's attempt to popularise the side engine position for buses had failed, the Q model of this layout being withdrawn from sale at the end of 1937, though its influence on other aspects of design was not dead, as will emerge later in this volume. On double-deckers – and indeed most heavy-duty single-deckers – the underslung worm-drive rear axle was used, with the worm casing offset to one side, generally the left (nearside), so that the transmission line was out of the way of the lower-deck centre gangway.

This layout had been adopted by G.J. Rackham for the design of the original Leyland Titan and, as well as being followed when he moved to AEC, had become all but universal on buses of that era. It had also become standard practice to make the bonnet, and hence the half-cab accommodating the driver alongside it, about as short as the engine would allow. At the other end of the bus, the rear entrance was still the overwhelming majority choice for double-deckers. There had recently been an upsurge of interest in the front entrance, not least in London Transport's country bus department, but this was beginning to wane. Among city fleets, the centre-gangway layout giving adequate headroom on both decks was almost universal. Although many provincial companies favoured the type with sunken offset gangway on the upper deck (generally known by what had been Leyland's trade name for the type – Lowbridge – because of its lower overall height) such buses were less numerous.

Within this remarkably standardised layout, there was widespread commonality of thought on the mechanical and structural specification. At that date, separate construction of body and chassis was still almost universal for all types of motor vehicle. Among the double-deck bus chassis makers, the oil (diesel) engine had become standard, even though still a rarity only five years or so earlier. Six-cylinder engines were in the majority, though the Gardner five-cylinder 5LW had built up quite a strong following in Daimler and Bristol buses. Engine sizes in common use ranged from 7 to about $8\frac{3}{4}$ litres and power output from 85 to 115 bhp, though there was if anything a trend towards the lower end of these scales. The debate on the pros and cons of direct versus indirect injection was still raging, with AEC as the main stronghold of the latter system and London Transport's STL class buses, in large-scale production at Southall, by far the most prominent application. Outside London, there was a clear swing towards direct injection, partly influenced by the exceptional fuel economy being obtained in fleets which had adopted the Gardner.

London Transport had inherited early experience with Gardner engines from the LGOC but had concluded that the limited speed range and other characteristics were unacceptable for its needs. However, there were clear indications that its engineers, in addition to the normal very close association with AEC on all aspects of bus engine design, were particularly interested at that stage in the products and engine policy of the other major British protagonist of the direct injection principle, Leyland. In this case the design adopted had a 'relaxed' characteristic, giving a relatively modest output in relation to engine size, but particularly smooth running towards the higher end of the speed range. Its fuel economy was not as good as that of the Gardner, but an important merit at that date was the consistent performance of the simple single-hole injectors it used, which were found to be less prone to the clogging then apt to be a problem with some systems using finer multi-hole designs.

AEC, conscious of strong competition from both Gardner and Leyland, and aware of the Swiss Saurer concern's introduction of such a system, had been developing its own ideas on direct-injection engines. An A171 7.7-litre engine had been converted from the standard Ricardo Comet indirect injection system to become the first prototype of this unit in direct injection form, being fitted in STL 709 in July 1935. The production version was designated A173, at first using quite a shallow cavity in the piston crown rather than the flower pot shape favoured by Leyland or the hemisphere of the Gardner design. Further examples of this were fitted to a handful of STL buses as well as some batches in provincial fleets.

Further work at AEC included experiments with both Gardner and Leyland combustion systems in AEC engines, but the A173 7.7-litre unit abandoned the shallow cavity with low cone-shaped base of the original 1936 production design for a toroidal (roughly doughnut-shaped) cavity during 1937. This was reminiscent of the 1934 Saurer dual-turbulence design, but the AEC version had relatively deep straight sides, and contemporary references to it were often to 'straight sided' toroidal. For the 8.8-litre, large enough to allow consideration of a design giving low output in relation to size, there was interest at both Chiswick and Southall in the Leyland-style pot cavity. Yet there was also a school of thought which still favoured the Ricardo Comet head, itself being improved, a Mark III version being standardised on STL buses from 1936. It is clear that the whole subject was still under review.

Transmission design was another subject showing quite wide variations of opinion. The fluid flywheel and Wilson-type preselective epicyclic gearbox were not only standardised in London Transport's fleet but fitted to all Daimler buses then in production as well as a sizeable proportion of the AEC buses supplied to fleets in the provinces for operation in urban areas, similar use being made of most of the Daimlers. The main challenger in providing an easier task for the driver than the traditional type of gearbox, in those days devoid of synchromesh on buses, was Leyland's Gearless system. This was an early application of a torque convertor and thus a predecessor of today's automatic transmission systems, though the driver used a conventional looking lever to put the unit into the direct drive condition, equivalent to top gear. This had found some favour in some urban fleets but most company undertakings and many municipalities still favoured the simple friction clutch and crash gearbox.

In other respects, however, there was wide-ranging uniformity. Almost all heavy-duty motor buses in Britain had vacuum-servo brakes, and AEC, Daimler and Leyland standardised on hydraulic systems to convey the driver's pedal pressure, magnified by the servo, to the brake drums at each wheel. The handbrake was always purely mechanical in those days, with a substantial lever in the cab. Although a few makes of car had independent suspension by that date, buses invariably had rigid axles and leaf springs. Double-decker, and most single-decker, chassis frames were of what could be described as the Rackham style, with the side-members shaped so as to permit a relatively low saloon floor level but gently arched over front and rear axles, generally sweeping down at the rear to support the entrance platform on rear-entrance double-deckers, though this last-mentioned portion was sometimes bolted on as a separate assembly. Because of the weight distribution, twin tyres were always used on each side of the axle at the rear, and this feature set a practical limit to frame width, which had to be reduced further at the front to allow an adequate steering lock for the front wheels – a particularly tight limit of a maximum turning circle of 56 ft between kerbs applied to London buses.

Thus in considering what form its future bus design might take, London Transport was virtually bound to follow many of the accepted standards set out in the preceding paragraphs, particularly since it had been decided that it would be a two-axle double-decker with engine at the front. Yet there was a strong desire to set the highest standards in terms of both visual and functional design. The LGOC had always been a design leader in terms of city bus development, and the standards of the still quite new London Transport had, if possible, to be even higher.

Beyond the prestige implications of a new type of bus for London service, the team of engineers had to address the practical needs for improvement. Operating conditions in London were becoming increasingly onerous in terms of traffic congestion, which meant that frequent use of the brakes was inevitable, to a greater degree for a bus than any other type of vehicle. Hence London bus brake performance had to be better than was regarded as acceptable in most other conditions. In this and other aspects of operation, it was not the very occasional 'crash' stop that was the problem in itself but ensuring that the brakes would stand up to repeated use without the need for frequent attention and maintain the ability to stop whenever the need arose, even after the most intense spell of 'normal' use. Similar remarks about durability applied to other mechanical features, but down the years it has often been the case that brake performance has tended to occupy the attention of London bus engineers more than any other practical aspect of design.

However, there was always a need to keep all types of operating cost down, among which fuel consumption and maintenance costs are key factors. Keeping the drivers happy was probably less obviously a primary aim, even though the end-result was to prove a major step forward in this direction. London Transport was large enough to employ highly-qualified experts in subjects not covered by smaller organisations, and valuable work was done on the best posture for the driver of a bus under urban conditions, for example. It was also regarded as important that a driver should not become unduly tired during a shift and thus more likely to become accident-prone in consequence. Trade union pressure was a factor in the consideration of this.

It is of interest to consider the people who were to play a major part in making the key decisions. As indicated in the introduction, A.A.M. Durrant, who had become Chief Engineer (Buses and Coaches) in 1935, taking in responsibility for Country Area vehicles as well as those of the Central Area, had a solid grounding of looking after the existing fleet dating back to the late 1920s before graduating to overall engineering control over London Transport motor buses and coaches. He and his staff, with several talented engineers among them, decided what was wanted.

As the agreement with AEC made by the LPTB provided that most of the Board's requirement for new buses would continue to be met by that manufacturer, it followed that the design team led by G.J. Rackham as AEC's Chief Engineer would be responsible for creating a chassis design in accordance with the LPTB's requirements. The principal people involved on both sides were well known to each other, and indeed one of the advantages of the close relationship between the LPTB and AEC was the detailed knowledge of each other's needs and capabilities. Most of the more modern vehicles in the LPTB fleet were based on chassis designs introduced by Rackham, but it was equally true that they had been extensively developed in regard to engines, transmissions, brakes and steering in response to needs shown up by London operating conditions on the basis of co-operation between engineers of both organisations.

Although the general proportions and layout of the new double-decker were to be much the same as the existing STL, there were to be significant changes in two key respects in terms of mechanical design, as well as a new much 'smoother' appearance. There was to be a larger-capacity engine and the adoption of an air-pressure system to operate not only the brakes but also the gear-change on the preselective gearbox. In addition, although many features of the chassis were similar to those of the previous model, the opportunity was taken to redesign the layout from the frame up.

On 6th December 1937 a model of the new bus was ready and shown to the Board's Engineering Committee, chaired by Frank Pick, Vice Chairman of the LPTB and a keen guardian of London Transport's growing reputation for good design. The committee asked Durrant to make a number of changes and the model was resubmitted early in January 1938. That the model was by now close in appearance to the final design is indicated by the purely cosmetic alterations requested at an Engineering Committee meeting on 10th January.

These alterations concerned the positioning of commercial advertising and the finer points of the livery. Regarding the latter, something a little different was obviously felt desirable for so modern a bus. At the beginning of 1938 the proposed livery differed from the standard treatment of the time in having two narrow silver bands above and below the lower deck windows. The meeting of 10th January decided that the double silver lines be replaced by single ones and that the black beading above and below the upper deck windows be changed to silver and continued round the front of the bus. A later change restored the double silver bands above the lower deck windows.

On 3rd February 1938 the Board asked Durrant to investigate the desirability of having route numbers placed on the nearside and offside top front corners of the bus in place of the central roofbox. He reported on the 14th that it was not considered practicable or desirable to put route numbers each side of the roof dome. Though no reasons are quoted in surviving documents, practical considerations must have included risks to a nearside roofbox from overhanging trees. Furthermore, it is difficult to envisage how such a design change could have been effected without harm to the appearance of the vehicle. Instead, Durrant proposed to the Board that the route number be shown on a stencil fixed below the front nearside canopy, and this proposal was carried forward to RT 1.

The credit for the sheer artistry of what was to prove one of the outstanding bus designs of all time is to some degree lost in the anonymity that usually applies to those who work for large organisations, but it is understood that a team led by Eric Ottaway was responsible and that Ottaway reported to Durrant. Their efforts, and presumably to some degree the personal 'eye' for proportions of Mr Ottaway, had been applied to some earlier London designs in which details that were later to be incorporated in the RT can be observed.

Facing page Even in the 1930s, operating conditions made heavy demands on the brakes of a London bus. Heavy traffic added to the frequency of stops and the unpredictable movements of other vehicles as well as pedestrians made the need for dependable stopping ability very important. The vacuum servo was beginning to reach the limits of its capabilities and something better was needed for bus operation in London. Buses themselves predominate in this scene in Oxford Street, though plenty of taxis are also present – note the rank in the centre of the street in the foreground. Most of the buses visible are of the STL type, which had become the largest London class by the time the design of the RT was being planned.
LT Museum

One important feature often credited to the RT as an innovation was not entirely new. This is the use of what is generally called four-bay construction, with four windows of greater than normal length on the lower deck. It had become common to provide either five or six equal-length bays in double-deck bodies and by the mid-1930s the five-bay body was the more usual (with the standard STL as an important example).

At the Commercial Motor Show held in the then new Earls Court exhibition building in November 1937, a strikingly fresh design of body on a standard 7.7-litre Regent chassis had been prominently displayed on the AEC stand. It was for Leeds City Transport and had bodywork by the local firm Charles H. Roe Ltd. Christened the Leeds City Pullman, it introduced the four-bay body to the contemporary double-decker. Numbered 400 in the Leeds fleet, it also had a well rounded profile with the upper deck front windows raked fairly strongly and the angle of slope diminished gradually to produce an almost continuous curve. The LPTB engineers seem bound to have been influenced to some degree by Leeds 400, but they took a quite different approach to window depth. The Leeds bus had unusually deep side windows, as implied by the use of the Pullman name, but the luxurious feel within was at the expense of a marked break in the continuity of waistline required to accommodate the standard Regent radiator height.

On the RT, comparatively shallow windows were adopted, amply big enough to give good vision for the seated passengers, but which combined with the low bonnet line to permit front and side windows on the lower deck to line up. Equally important in terms of the visual design, the bonnet level aligned well with the body waistline.

Above left Double-deckers of uniform size and general layout were to be found in the fleets of most major operators in Britain in 1937, when design work began on the RT. The standard London bus was then the STL, based on the AEC Regent chassis with 16ft 3in wheelbase, which had evolved into the form shown by 1936, with A171 7.7-litre engine, fluid flywheel, Wilson-type preselective gearbox and vacuum-hydraulic brakes. The 56-seat timber-framed body, built by the LPTB at Chiswick, was to an outline introduced in November 1934, and STL 1301 is seen when new early in 1936. The wheel discs were not standard, though the type fitted on the rear wheels was to be standard for the RT. *LT Museum*

Left Leeds City Transport's No. 400 was exhibited on the AEC stand at the Commercial Motor Show held from 4th-13th November 1937, and its probable influence on the RT body design is not hard to detect. AEC planned its show exhibits over six months in advance and the general form of this vehicle would have been known to Rackham and, very probably, LPTB designers well before the event. The body was designed and built by Charles H. Roe Ltd, the Leeds undertaking's main body supplier. Roe had begun to standardise on the curved profile from 1936, but the use of four-bay layout between the front and rear bulkheads of a double-decker of otherwise orthodox layout was a new feature.

In spring 1938 Ottaway gave a talk to the Design and Industries Association in which he spoke on the subject of bus design. He pointed out that while the value of streamlining on fast moving vehicles was considerable, its practical advantages when applied to slower types of transport were "relatively small and limited". STL 857 (later STF 1) had been built in 1935 with a streamlined full-fronted body but was converted to half-cab in May 1938. "An exaggerated streamline", commented Ottaway, "is inappropriate to a public service vehicle, which should, however, have rounded corners and a continuous and smooth contour, with an emphasis on the horizontal lines". He believed that a bus should encourage riders by its attractive appearance.

Reflecting recent advances in interior bus design, he went on to point out that the "same care should be bestowed on the lines and proportions of the interior as that given to the exterior". He pointed out that careful selection of colours could create a cheerful and restful impression, and said that Rexine-covered pressed aluminium finishers should be used to eliminate dust traps.

There was in fact little that was revolutionary about the interior of the RT; the revolution had taken place a few years earlier with the 5Q5 single-deckers of 1935. Many features of these (also used on subsequent single-deckers) were simply transferred to the RT with little or no alteration; the aluminium framed seats, the gently curved corners of the windows and the extensive use of easy-to-clean Rexine all made the move highly successfully.

At first the new project was simply referred to as the "New Double Deck Bus" but by July 1938 the code RT was being used. The Engineering Committee minutes of 25th July state that "In view of the long time required for delivery, Mr Durrant asked authority to place an order forthwith with the Associated Equipment Company for 150 chassis for 'RT' double-deck buses for delivery during the (financial) year 1939/40".

Above right Thoughts on bus design two years before work started on the RT. The experimental STL 857 followed a mid-thirties fashion for 'streamlining' but was not liked by staff. Chiswick's designers possibly had in mind what might come after the STL in terms of body design, but they soon had second thoughts about this one and the vehicle was converted to half-cab in 1938. *LT Museum*

Right This view of one of the Q-type Central Area single-deck buses of 1936 shows the use of pressings with a rounded section and radiused corners as internal window finishers, in much the same style as later adopted for the RT. This method may have been influenced by Park Royal Coachworks, which built these bodies and which used similar rounded section on its own metal-framed designs of the period, though not then with radiused corners. Also evident is London Transport's distinctive design of tubular-framed seat with double top rail, then newly-introduced and standard for LPTB buses generally from that date – nearly 60 years later, many designers of modern buses would do well to study its level of comfort. *Park Royal*

The oil engine was firmly established but the jury was still out on many aspects of its design. In a general way, both in regard to petrol and oil engines there had been a long-established trend of extracting more power from engines which often were slightly smaller than their predecessors, model for model. This was often a natural by-product of technological development, but unless all aspects of design kept pace, there could be a drop in durability. London Transport was finding that the A171 7.7-litre oil engine being used in the STL was proving less durable than the 8.8-litre unit that had been adopted for the LT-class six-wheel double-deckers on AEC Renown chassis, of which over 400 were then running, mainly with A165 engines and many as conversions from petrol.

In addition, there was the question of the form of injection system, as already mentioned. The Ricardo indirect-injection engine gave good power output but was less efficient in terms of fuel consumption than direct-injection designs, as well as being less easy to start from cold, with a need for the complexity of heater plugs. London Transport seems to have been less troubled by this last characteristic than some other users of AEC-Ricardo engines of that period, perhaps due to London's climate being more benign than is the case further north and also to the development of suitable lubricating oil, in which LT chemists played a significant part. There was something of a battle of wills over this topic both at AEC and within LT, and it is significant that the latter seems to have taken a keen interest in Leyland direct-injection engines in buses taken over from independent operators.

Indication of London Transport's satisfaction with the Leyland engine came with the order for 100 Leyland Titan buses based on the contemporary TD4 design which became the STD class, delivered in the spring of 1937. The design was modified quite extensively to meet LPTB requirements, this applying to chassis details as well as the Leyland-built bodywork which was given an appearance superficially akin to the contemporary STL, but the 8.6-litre engine was virtually a standard Leyland product. At the time, the LPTB engineering staff were struggling to re-establish their own standardisation and getting rid of most of the non-standard models acquired from independent operators as quickly as possible.

So the introduction of an unfamiliar type was not being done lightly – doubtless, there may also have been an element of quite deliberately farming out part of the Board's requirement for new buses to keep AEC on its toes. But the evidence was soon underlined that Leyland's engine, with its low output in relation to size but good durability and particularly smooth running towards the top end of its speed range, was influencing the line of thought among LPTB's engineers.

The whole question of suitable engine size and design was under review by both LPTB and AEC engineers. The latter concern was giving the impression in its publicity material for that period that it had standardised on the 7.7-litre engine for all but its lightest models, but in fact was meeting a continuing demand for larger engines from some users of eight-wheeled goods vehicles and the Great Western Railway, to whom it was supplying diesel railcars, as well as for Regent double-deck buses for several municipal operators in hilly areas. This was being met by continued production of the 8.8-litre engine, initially in A165 Ricardo Comet head form, much as had been produced since 1932.

A conversion of the A165 engine to direct injection, using the same form of shallow cavity in the piston as in the early version of the A173 direct-injection 7.7-litre engine, had been produced by February 1937 when a converted engine (A165J-850) was placed in service in LT 1371, one of the 'Bluebird' 60-seat AEC Renown buses that had had A165 engines and preselective transmission since entering service in 1932.

However, more significantly, another 8.8-litre engine was running on test at AEC converted to have the Leyland pot-cavity type of direct injection, a March 1937 experimental department report speaking of "very encouraging results". Usually such an engine would have been considered no more than a test unit built for comparative purposes, but new engines to similar design were to enter service in London in significant numbers from the following year.

It is clear that by the summer of 1937, there was a strong revival of interest at both Chiswick and Southall in engines of the 8.8-litre class. At AEC especially there was also no doubt acute awareness of the fleet of 100 STD-class Leylands, with their smooth-running engines of almost similar size, by then operating on London's streets.

The decision was made to set about the design of a completely new engine, having little more than the 115mm bore and 142mm stroke in common with the A165 and its predecessors. It was given the project number A805/1 and a report dated 27th October 1937 describes the initial bench tests of the first hand-built engine. A main aim was to improve durability by providing more bearing area for the crankshaft, itself strengthened, and the distance between the cylinders was widened slightly. This latter made allowance for an increase in bore size to 120mm, hence giving the 9.6-litre capacity later to become familiar, though this was not mentioned in the report. Gear instead of chain drive was adopted for the timing and tidier front-end design reduced the overall length of the engine, despite the wider spacing of its cylinders, to a point that made it only marginally longer in terms of installation than a 7.7 litre unit.

Another feature of the A805/1 engine design was the capability to allow a number of different combustion chamber designs to be incorporated. It is very surprising, with hindsight, to discover that this first engine was of indirect-injection type, the Ricardo Comet III* type (the asterisk forming part of the designation). AEC's eminent chief engine designer, C. B. Dicksee, is known to have been 'pro-Ricardo' and evidently won the day for this first version of the new engine. In terms of power output, the results were impressive, for with maximum

Among the Leyland double-deckers (mainly Titans) acquired from independent operators after 1933, London Transport inherited five with Leyland oil engines. Significantly, this number was expanded to nine by four more conversions from petrol carried out by the LPTB; they ran until general withdrawal of the TD class in 1938-9. The choice of Leyland as supplier of 100 new Titan double-deckers based on the TD4 model in 1937 was a further indication of LPTB interest in this concern's products. Various non-standard chassis and body features were incorporated but the 8.6-litre engines were virtually standard – although no specific maximum power figure was given for the London examples, the standard output was being quoted at about 94bhp at the time. Particularly smooth running towards the top end of the speed range was characteristic of this engine. Here STD 90 is seen before delivery from Leyland – the destination blinds were probably no more than a set used for photographic purposes. *Leyland*

fuelling it gave 145bhp at about 2,050rpm, and even though this would doubtless have been cut back for normal service (10 per cent derating would still have given 130bhp), it is a remarkable figure for that period, and indeed more than was wanted at the time, except in very exceptional circumstances. The greater fuel economy of the direct-injection system was to cause effort to be concentrated on engines of this type and, initially at least, the Leyland pot-type version was strongly in favour with London Transport despite much lower power capability, and indeed here the prospect of a large-capacity yet compact engine was attractive.

However, this new design of engine was not yet ready for production and it was the interim A165-based pot-cavity engine (this feature being incorporated under licence from Leyland) that was chosen for a new fleet of coaches to be built for London Transport in 1938. They were for the Green Line limited-stop services running across London in most directions, which had been established with petrol-engined AEC Regal models numbered in the T-class and dating from 1930-31. These were due for replacement and some 266 oil-engined

Regals were ordered accordingly. Visually, they had bodywork similar to that on a batch of 50 placed in service in 1937, generally identified by the engineering code 9T9, but the new 10T10 batch of 1938 switched from the A171 7.7-litre engine to a pot-cavity 8.8-litre unit, initially designated A165Z though soon becoming A180. Some effort was made to reduce the length of the installation without attempting anything like the A805/1 redesign and these vehicles had no fan (possible because of the direct injection engine's higher efficiency) and a repositioned water pump to allow the radiator to fit very snugly in front of the engine timing case.

The 10T10 coaches soon earned a reputation for refined running as well as general capability, becoming popular with passengers, drivers and engineering staff, and reinforcing the trend of thought towards the use in London Transport double-deckers of engines having similar characteristics. In addition, a further 550 A180 pot-cavity engines were built in 1939-40 for conversion of most of the LT class double-deckers still fitted with petrol engines, and others were supplied in new Regent buses outside London.

Meanwhile engine number 3 from the A805/1 experimental batch was selected for fitting to the prototype RT chassis (although not yet known as such) that was under construction at AEC in the spring of 1938, which thus began as an 8.8-litre vehicle. AEC records show the full number of the original engine fitted in this chassis as LH A805/1-3, and the significance of the LH prefix is a matter for speculation. 'Left-hand' obviously jumps to mind, but makes no sense in regard to this vehicle – to AEC at that date, a left-hand engine meant one with auxiliaries on the opposite side to standard, such units being built in small numbers for marine or industrial applications. In later years, true left-hand engines were built by AEC for left-hand drive forward-control chassis as exported in large numbers to South America in particular. Another explanation, much more plausible to the author, is that LH signified 'Leyland head', just the sort of shorthand apt to be used in an engineering department to signify the use of the Leyland type of combustion system in this version of the new engine, particularly as the first engine had a different form of cylinder head.

In March 1938, three more 8.8-litre engines of the set built as part of the A805/1 project, serial numbers 5, 6 and 7, were fitted in the last three of a batch of 327 STL-type buses then in course of delivery and ordered at the same time as the first batch of 10T10 coaches. They had chassis O6615940-42 (the O standing for oil-engined) and were numbered STL 2513-2515, receiving what were evidently standard STL12-type bodies. No photographs of these buses in original condition appear to have been taken, but all the evidence suggests that they would have differed little from the rest of the batch in external appearance. In particular, the bonnet length was probably no greater than standard, or so little as to allow the standard bodies to be fitted without infringing the 26ft overall length limit of the time. They were allocated to Hanwell garage (in later years known as Southall), frequently the home of experimental buses and conveniently close to the AEC works. They were thus an important stepping stone to the RT and may well have sounded quite like an early RT in internal noise level and tone. In 1942-43, they were converted to standard, becoming indistinguishable from the rest of the batch.

No further information on these A805/1 engines has come to light, but another candidate for experimental running with one is an AEC Regent chassis, O6615413, which, fitted with a Weymann bus body, was registered by AEC in Middlesex as JML409 and placed in service as a demonstrator with Bradford Corporation on 10th November 1938, by whom it was purchased in 1940. Its engine number appears in Bradford records as U101898, this also being in the U series used for all kinds of experimental department items, and sometimes used to provide a form of unit or chassis serial number – it seems possible that in this case the rather complex A805/1 serial numbers, though acceptable within AEC or the closely allied London Transport, were thought unsatisfactory for a vehicle going to an outside organisation.

It seems very likely that this chassis may have been used by AEC for road test running of the new type of engine before being sent for bodying. The appearance was almost indistinguishable from that of a 7.7-litre Regent, with only the lack of a starting handle shaft as a clue to a flexibly-mounted engine, and hence something 'different'.

Meanwhile, the prototype RT chassis, numbered O6616749, was being completed and tested, being delivered from AEC to Chiswick on 23rd May 1938, officially becoming LPTB property on 30th June. The latter's stock card, at first using the temporary fleet number ST 1140 as explained later in this chapter, shows the bore and stroke as 115mm and 142mm, the 8.8-litre dimensions. No later change to this appears on the card even after the vehicle became RT 1, although it was stated to be of 120mm bore and 9.6-litre capacity in published reports clearly based on official information when the complete bus was shown to the press in July 1939. It is known that no 9.6-litre engine to the new design was in existence up to 8th December 1938, nor seems likely to have been for a further month or two, as an AEC experimental department report of that date refers to preliminary tests being carried on what was described as "a present-standard 8.8" almost certainly an A180 – with bore opened out to 120mm, "to obtain advance information regarding the performance of the new design 9.6-litre engines to be built next year". These latter were almost certainly primarily for the RT but possibly also for railcar application.

Although many of its detail features had a familiar look, the RT chassis was fundamentally new, every item having been reappraised and freshly designed where this was justified. This was true of the frame, which differed from the standard Regent unit of the time in having a parallel section at the front, instead of tapering outwards from the front dumb-irons and reaching full width at a point under the front passenger seats.

The RT frame was also tapered outwards, but beginning roughly where the earlier version ended and reaching full width almost half way along its length. This had the effect of allowing a little more room for the tight steering lock angle needed for the front wheels, but there were also other changes, such as the steeper angle of the front dumb-irons allowing for a simple way of providing radiator mountings.

On the original Regent design, the radiator was carried by the front of the engine, but this was not compatible with a flexible engine mounting, another important new feature of the RT. It could not claim to be the first such system on a bus chassis, for Daimler had been using quite an effective one for a couple of years or so, but the RT employed a particularly neat design. What were virtually two rubber rings, one high up at the front of the engine and the other quite a large unit behind the fluid flywheel, were each inclined so as to allow the engine assembly to oscillate about the axis between them, this being designed to pass through the unit's centre of gravity. As a result, noise and vibration within the bus were much reduced as compared to the contemporary standard STL.

The main new feature of the chassis was undoubtedly the air-pressure system which operated both the brakes and gearchange – it also supplied a convenient means of operating an automatic lubrication system.

The circuit diagram for this was reproduced when details of the new model were released to the technical press, and at the time seemed very complicated. It took a long time for the bus industry outside London to realise the virtues of such a system, yet gradually, as traffic congestion in provincial cities became more nearly equivalent to that in the capital, the principles were accepted and, indeed, many of the buses in service today have systems which are very similar in basic concept to that of the RT.

Air pressure has a fundamental advantage over vacuum as a means of reducing the amount of effort required from the driver to operate the brakes on a bus. The normal atmospheric pressure of about 14.7 lb/sq in means that effective vacuum 'pressure' cannot be more than that. However, with a compressor, an air pressure of 100 lb/sq in is readily attainable, meaning that the required rate of braking became more easily attainable, with a reserve of capability, and the size of the operating cylinders could remain modest.

Trolleybuses had generally adopted air-pressure brakes by the early 1930s, and they had been fairly common on various makes of early six-wheel motor bus in the late 1920s but, in Britain at least, had been discarded in favour of the Dewandre vacuum-servo brake system. The problem at that date seems to have been difficulty in providing graduated control, particularly important on a bus, and in which respect the Dewandre system worked well.

Vacuum was readily obtainable from the induction manifold of a petrol engine and thus vacuum-servo brakes had become the accepted choice for motor buses. On a trolleybus, no such vacuum source was available, and this, together with often greater weight, may have helped to tip the scales towards air, and by the time the trolleybus was becoming popular in the early 1930s, it seems that the controllability problem was being overcome.

The prototype RT chassis at Southall in March 1939 had completed its period of service with temporary body as ST 1140 and various new items had been fitted, including the radiator, slightly different from that used originally and setting the standard for subsequent production RT buses. It is noteworthy that the frame still has the rear extension, which would have been essential for use with the open-staircase body but which was presumably cut off after arrival at Chiswick to suit the new body. By this stage, it is believed that its engine was to A185 specification and with 120mm cylinder bore, giving the 9.6-litre capacity, though whether it was technically the original A805/1 unit modified or had a new identity, still almost certainly with an experimental unit number, is not known – the unit's appearance conforms to the production version except in minor details. *AEC*

Bottom left This view, also dating from March 1939, as well as showing part of the engine installation in the prototype chassis, also illustrates the tight steering lock required to meet London requirements, the achievement of which was made easier by the parallel-sided frame at this point. The air brake cylinder on top of the axle king pin was different in design to the production unit used for the 2RT version, and thus one of the items that may have led to the decision to dismantle this chassis in 1946. *AEC*

Bottom right The air-pressure system was a major departure from previous practice, incorporating as it did the brake system, the means of operating the pre-selective gearbox, visible on the left of this view of the chassis, and an automatic lubrication system. The set of small-bore pipes for the last-mentioned can be seen clipped in place across the frame. The fluid flywheel is just visible in the housing behind the engine. *AEC*

There seems little doubt that the standardisation of air brakes on the hundreds of trolleybuses entering service in London helped to underline the potential. Even though the tram and trolleybus department of London Transport was quite separate, with its own engineering team at Charlton works working completely independently from the bus engineers at Chiswick, it would have required remarkably blinkered vision for the one to be unaware of the other's activities. This was especially so in regard to brakes, perhaps the main preoccupation of anyone concerned with the safety and maintenance of a fleet of buses with either type of motive power working in London's traffic.

Hence, much of the technology was readily available, and indeed air-braked heavy goods vehicles were already well-established in some countries. A system using an air cylinder for the brakes on each wheel was adopted and, thus far, the design was based on existing ideas, even if rare in terms of British motor bus practice. The control valve was designed so as to give progressive braking in proportion to pedal pressure, but with a deliberate heavier action beyond a level regarded as 'normal', discouraging use of maximum braking in other than an emergency.

A significant step forward was the realisation that having a supply of compressed air on the vehicle provided a solution to a drawback of the Wilson-type epicyclic gearbox in the form in which it was then being used. This unit relieved the driver from repeatedly having to press and hold down what was inevitably a 'heavy' clutch pedal, as required with a conventional gearbox when in congested traffic. However, the use of the equivalent pedal to effect the gear change within a Wilson box, after the choice of gear had been preselected by the lever, was still quite hard work.

Anyone who has driven a bus or even a car with this form of transmission in its original form will know that the pedal action, as well as being quite heavy, had a curious 'feel', the pressure needed to move it being greater with the pedal 'up' than 'down'. This was due to the nature of the internal mechanism which maximised the leverage of a strong spring to give the grip needed on the friction bands for each gear.

There had been realisation in the early days of the use of Wilson gearboxes at AEC that this was a drawback, a vacuum cylinder being used on some early Q-type buses to operate the change, acting against the internal spring. A similar idea, but using air pressure, was tried on a new underfloor-engined single-deck model, the TF, being developed jointly by London Transport and Leyland, the first example of which was just a little ahead of the RT in its development stage. This had the disadvantage that if the vehicle had been left in gear when parked, quite possible with a fluid flywheel, it might not be possible to engage neutral until after the engine had been run and built up enough air pressure to operate the gearbox.

However, for the RT, the principle used was different in that the air pressure was used directly to replace the internal spring in the gearbox, being supplied via a pedal-controlled valve, which thus engaged the gears after they had been preselected by use of the hand control. This meant that any loss of adequate air pressure if the vehicle had been out of use for a time would mean that the gearbox would remain in neutral until pressure had built up, thus ensuring that it could not be driven until the brakes also had enough air pressure to operate satisfactorily. More significantly, in normal use, air-pressure operation meant that both brake and gear-change pedals could be made quite light in action, thus greatly reducing the physical effort during each spell of driving in London traffic.

Another change was the removal of the pre-selector lever from cab floor level as on the STL (where it looked very like a conventional gear lever), to the steering column. It was turned on its side and made quite short, but still operated in an H-pattern like a normal gear lever and was placed so as to be just under the driver's left hand when in its normal position on the steering wheel. The idea of a control on the steering column was by no means new, and most of the early users of Wilson gearboxes favoured a quadrant control so placed. The AEC system, both in floor and column form, had the virtue of eliminating uncertainty as to which gear had been preselected, which was a problem with the quadrant system, particularly in the dark.

The idea of a steering-column control for a synchromesh gearbox was beginning to come into favour on American cars, and the similarity of the RT layout to this was mentioned in early publicity. It may have reflected G.J. Rackham's ongoing contact with transatlantic ideas, for he continued to make frequent visits there after his spell with Yellow in the 1920s. In fact, the RT preselector version worked very much better, not having to convey the far heavier loading involved in moving the synchromesh mechanism for each gear in such a box.

The internal mechanism of the gearbox, which AEC designated D140 in the production form with air operation used for the early RT chassis as built in 1939-40, was very like that in the previous D132 unit used in the London STL and other fluid-transmission AEC models of the period. The set of compound epicyclic gears was basically unchanged, giving rise to the familiar type of whine in the indirect gears, and indeed in neutral if this was used while at rest, though the RT added a characteristic 'hiss-clonk' as a gear was engaged.

As to the other driver's controls, the steering was basically of the standard AEC worm-and-nut type, as originally evolved in 1931 to meet London requirements. This used lower gearing than generally favoured elsewhere and, combined with the choice of relatively small-section high-pressure tyres on the front wheels, quite light action resulted, even if the driver did have to turn the wheel through a greater angle for a given corner. In those days, long before power-assisted steering was developed, it was a well-judged design, and particularly smooth in action when correctly set up. For the RT, the steering column angle was made more nearly vertical than on the STL and standard AEC passenger models of the time, this being in response to a study of the most desirable posture for the driver carried out by London Transport's medical staff.

The other sizeable control in the cab was the handbrake, continuing from previous models as the long lever to the right of the driver, though a refinement on the prototype and early production models as originally built was the inclusion of an oil bath for the ratchet mechanism to reduce the wear from frequent use.

An unfamiliar feature associated with the change to air pressure was the small red 'Stop' arm which was placed above the windscreen to drop into the driver's line of vision if the air pressure fell below an acceptable figure – it was generally called a flag, though the effect was more akin to a miniature semaphore signal.

The front axle was altered to bring the centre point around which the road wheel swivelled when steering as near as possible to the centre of the area of tyre in contact with the road, to lighten the steering. The rear axle had a heavier-duty worm drive, but the ratio on the prototype and early production RT remained unchanged from the 5.75 to 1 of the standard STL type.

The most immediately obvious feature of the new design was the lower mounting adopted for the radiator. At the time, this seemed quite surprising, as the tendency had been for radiator height to increase somewhat on most types of vehicle during the period from the late 1920s – earlier models of car as well as bus with low-mounted radiators were apt to be regarded as old-fashioned. Yet, from a practical viewpoint, especially when driving in heavy traffic and repeatedly having to pull out from behind other vehicles, it made sense for drivers to be able to see directly just how much space there was in front of the nearside mudguard. The low bonnet line helped in this respect as well as giving a tidy appearance by matching neatly with the body waistline, though this latter did not show up until the chassis was united with the equally new design of body intended for it. However, the decision to go for a design which broke with accepted practice was an illustration of how London Transport, already well established as a design leader as well as big enough to ignore convention if it seemed justified, could set rather than follow fashion.

There were only a couple of inches or so of clearance above the top of the engine to the bonnet top, but the concept had a logic about it that was typical of London Transport's blend of advanced ideas and practicality, apt to be taken up later by others. The war delayed matters but in the immediate post-war period from 1946, Bristol, Crossley and Guy all adopted similarly low-mounted radiators, though AEC, ironically, reverted to 'normal' for its post-war Mark III versions of passenger models derived from the RT but intended for sale to operators outside London. This makes it clear that the low radiator idea came from Chiswick rather than Southall.

As mentioned earlier, the automatic lubrication system was another feature operated from the air pressure circuit. This was to a design developed by two engineers on the staff of Birmingham City Transport named Rowlands and Parker and marketed under the name RP via the Clayton Dewandre concern. It fed oil to 24 points around the chassis, covering the spring shackle pins, the front axle king pins and various items related to the mechanical aspects of the air system. This reduced the need for maintenance by eliminating the need for regular greasing of many of these items. Another item based on a similar philosophy was the RP automatic brake adjuster, also adopted for the RT.

Superficially appearing to be an 'old bus', the prototype RT chassis is seen here with the body removed from TD 111, an ex-City Leyland dating from 1931. In this photograph, the fleet number ST 1140 had yet to be applied, and a trade plate, 002 GH, was being used, suggesting that the vehicle was being taken for initial trial runs in July 1938 before its period of service. At this stage, the part of the radiator visible below the frame was thicker than the later version. Note the simple style of cab constructed at Chiswick to adapt the body to this chassis and how the lifeguard rails have been cut to suit part of the air system, itself differently arranged as compared with that shown in chassis views the following March, the air tank in this earlier picture being further forward. City's body style differed from the normal version built for London operators by Dodson, the front of the upper deck being rather reminiscent of the LGOC NS type. The unladen weight of the vehicle was 6 tons 14 cwt 1 qr.

Clearly, London Transport's engineers, and indeed no doubt also those at AEC, wanted to see how the new bus would behave, but the new body was not yet ready when the chassis arrived at Chiswick in May 1938. It seems in any case that the engineers wanted to try out the chassis secretly before publicising the new type to the wider world. It might have been expected that a standard STL body would have been modified – by no means an unduly difficult task for Chiswick's resourceful bodybuilding department, for AEC's designers had done a good job in keeping the RT bonnet length down to only 1³⁄₈in more than the STL. Such a plan might have given a good degree of anonymity, for the result would have looked like an STL with a low bonnet – odd-seeming, but not unduly unusual when experimental modifications were quite often to be seen among vehicles entering service.

Instead, it was decided to go in for what seems as a quite amazing 'cloak and dagger' type of disguise. The bus was to be given the temporary fleet number ST 1140, at the end of the ST series of 15ft 6½in wheelbase Regent chassis, most of which dated from 1930-31. Adding the new vehicle to this series ignored the fact that the STL code signified the longer 16ft 3in wheelbase chassis standard from late 1932, and that its chassis was marginally longer still, at 16ft 4in.

To complete the 'old bus' illusion, the body selected for this state-of-the-art chassis, bristling with modern features, was an open-staircase one removed from a TD-class Leyland, this type by then being in course of withdrawal. By 1938, double-deckers of such types were regarded as very out-of-date, for the

previous decade had been one of remarkably rapid development. The donor bus was TD 111, which had entered service with the City Motor Omnibus Co Ltd in 1931, one of a batch of six vehicles which were officially recorded as Tiger chassis of type TS3, and having chassis numbers in the appropriate series (this one being 61748). Some Leyland records quote them as 'TD1 special', which gives the clue to their nature, for the Tiger TS3 and Titan TD1 were essentially similar, both with a 16ft 6in wheelbase.

The body from TD 111 had been built to the City company's requirements by Dodson, responsible for bodying many of the London independent operators' buses, and seated 56 (30 upstairs, 26 down), which was more than usual for a two-axle design of that period. To suit the RT chassis design, the cab was modified. Indeed, various other changes would have been needed, such as a floor trap to give access to the amidships-mounted gearbox, that on the original chassis having been just behind the engine.

Records show that the new chassis (whose number O6616749 was in the normal Regent series, incidentally, following on from a further batch of STL buses then on order) became LPTB property on 30th June 1938. The ex-TD body was mounted the following day, 1st July, so no time was wasted. It received a PSV licence on the 9th and was allocated to Hanwell (HW) garage on the 13th of that month – as mentioned earlier, that location was often used for experimental buses, being only a few hundred yards from the AEC works. The London Transport Allocation Book for August 1938 had an entry reading "An experimental vehicle known as RT 1 occasionally operates on service 18c from Hanwell garage", indicating that some members of the operating staff were already referring to the prototype by that number, though that may have been no more than an unintentional revelation of the number it was to receive with the body meant for it.

Just what public reaction there was to the appearance of this rather odd-seeming 'old' bus has evidently escaped being recorded. Route 18c ran from the garage to Wembley, Empire Pool. The more observant would notice that it had an up-to-date registration number, EYK396 – others in the EYK series having been allocated to the latest batch of Green Line coaches. Doubtless it sounded somewhat like them, too, with the combination of an 8.8-litre pot-cavity engine with AEC-built preselective gearbox, though the engine to A182 pattern then fitted, plus the details of its installation, may have given a pretty close approach to the rather deeper note later to become familiar on the earlier production RT buses.

The technical press seems not to have spotted it, or if it did, not to have reacted – the relationship with manufacturers as advertisers was apt to be more cautious in those days than has been so in more recent times. After 5½ months, the experiment was deemed to have served its purpose. ST 1140 was delicensed and the body removed on the last day of 1938, though the latter was not sold for scrap until 12th April 1939.

Then began the transformation into what had been intended from the start – a new London bus that would set the pattern for years to come, for far longer as it turned out, than any of those responsible could have foreseen. Although the running in the guise of ST 1140 had confirmed that the concept was fundamentally on the right track, there was some work to be done on the chassis. It returned to AEC – the exact date of this is not known, but doubtless soon after the ST 1140 body had been removed – and what appears to have been quite extensive work was carried out on it, as conveyed by a series of photographs taken after this process was completed. Much had been given a fresh coat of paint, but a number of items, including the engine, show evidence of having been newly fitted, with items in metals such as aluminium or brass unpainted and too clean to have run for even a few months.

Although the record card shows no change of engine at this point, it seems clear that by that stage the unit was to the specification later known as A185, with 120mm bore and 142mm stroke and hence the 9.6-litre capacity then so new and remarkable when the trend had been to smaller engines. The identity of the engine at this stage is not clear – it may be that, technically, it was regarded as a rebuild of the original unit. This is possible in practical terms, though it seems more likely that a completely new engine to the 9.6-litre specification would have been built and run on the test bed to ascertain optimum settings before installation.

Other changes included a revision to the radiator outline at the bottom, again to the form that was to become standard for subsequent London RT buses and the relatively small numbers of this type that were built for other operators.

After the old body was removed, the chassis was returned to AEC and it is clear that quite extensive modifications were made before it was ready to be sent to Chiswick again to receive the new body and become RT 1. This is a photograph taken at that stage, the negative being dated 6th March 1939. The air-pressure equipment had been altered and the large gauge mounted on the temporary dash panel allowed checks on its performance to be made, implying that further testing was being carried out. Apart from various new items, the chassis had been generally tidied up, but a clue to the fact that it had already been in service was the painting of the front hub cap, typical Chiswick rather than Southall practice. It is interesting to note that the chassis frame then still had the low-level rear section to support the entrance platform of the body, as was usual on double-deckers generally and had been essential when carrying the open-staircase body as ST 1140. *AEC*

The new body was not only dramatically new in its appearance but was of quite different construction to other Chiswick products of the time. The standard STL body, though quite modern-looking, was of timber-framed construction, using mainly ash and designed for a life of about 10 years with the expectation that it would be given very thorough intermediate overhauls, being removed from the chassis for the purpose. In earlier periods, this had been an annual procedure, and although this period was beginning to be extended, the idea of expecting to have to carry out quite a lot of work keeping bodywork up to a high standard had been generally accepted.

A statement on the RT's costs made to the Board by Durrant on 3rd February 1938, at an early stage in the development of the bus, had quoted an estimated extra £130 per vehicle compared with the latest STLs (£1630 against £1500, an increase of around 9%). £80 of this increased cost was attributed to the body in respect of improvements in construction, finish and appearance and the cab. The financial forecast showed that the benefits in running the new buses and the extended life of the body were expected to give a net annual cost saving of about 15%.

The Board did have some batches of metal-framed bus bodies, and this form of construction was standard for the fast-growing trolleybus fleet, but these had come from other bodybuilding concerns. Some were beginning to prove their longevity and freedom from the need for frequent overhaul. It is clear that the body department at Chiswick intended to follow a similar path, and indeed had begun to do so to some degree.

The phrase 'metal-framed', if not as misleading as 'all-metal' in relation to body construction, is one that should be treated with caution, for many so-called metal-framed bodies of that period included quite substantial amounts of wood – notably as a packing material within pillars whose main function was often the attachment of the external panelling, the latter nearly always in aluminium alloy. It had been planned that the bodywork on the 10T10 Green Line coaches of 1938 would be 'all-metal' but difficulty in getting supplies of sheet steel was quoted in a report to the Board by Durrant in May 1937 as the reason for adopting an improved form of composite construction for them with flitched wooden cross bars and pillars and metal longitudinal rails.

This page RT 1 just after receiving its intended bodywork on 27th March 1939, seen outside the Experimental Shop at Chiswick. The registration number had yet to be painted on the panel incorporated at the foot of the radiator and the unladen weight figures were yet to be added. The hint of what might have been regarded as 'vulgarity', in the polished aluminium mouldings, was soon to disappear. Even so, the sheer artistry of what was to prove one of the classic bus designs of all time was already evident.

Facing page The offside view taken at the same time. The livery, which was to have such a short life on the completed vehicle, had been agreed over a year earlier at a meeting on 10th February 1938 just after the exterior design of the bus had been completed in model form. It was a revision of an earlier scheme which would have used two polished bands below the lower deck windows as well as above. It seems probable that in this earlier plan the two lower bands would have merged at the front offside corner of the cab.

The body for RT 1 was designed to develop further what had been learned and also to take advantage of the rigidity of a well-designed metal-framed structure. A double-deck bus body has great stiffness if suitably constructed, and the idea of suspending the rear platform from the structure above was not only logical in structural terms but allowed the chassis to end just behind the rear body mounting points under the bulkhead. This meant that a rear-end collision would hardly ever cause damage to the chassis frame, as was liable to be the case with the usual design at the time.

The staircase design was altered to include an extra step, on the logical premise that shallower steps are easier to negotiate, even though there were now eight rather than seven. As a result, the bottom staircase step was moved $4\frac{1}{2}$in nearer the platform edge than on the STL. The internal saloon heights for the prototype are quoted as 5ft $10\frac{1}{4}$in for the lower deck and 5ft $8\frac{1}{4}$in for the upper deck.

Another neat detail was the way in which the problem of how to deal with the awkward-looking corner behind the near-side front mudguard was overcome. This was also a feature which had been tried out on other models, including one double-decker, STL 2434, as well as the 10T10 coaches. In effect, the mudguard was carried by the bodywork behind it, which meant that it could be neatly faired-in, the front half being quite easily removable for maintenance access.

For some unknown reason, the rear of the RT did not match the curvaceous front. On the STL, and most other contemporary British double-deckers, there was a curved rear profile, though the degree of this was much less than at the front. On the RT, the rear panels were vertical all the way up to the upper-deck waist and though the rear dome, incorporating a neatly-finished single-window emergency exit, was well rounded, the overall effect gave a slightly tail-heavy look.

Though the driver's cab remained very plain-looking internally, two significant changes materially improved driver comfort, quite apart from the attention paid to reducing the physical effort involved in driving. Hitherto, LPTB buses had followed LGOC practice in not having cab doors, evidently a legacy from the spartan regime imposed by the attitude of the Metropolitan Police when in charge of vehicle licensing. However, there was also a practical problem in that buses were parked up very closely alongside each other in some London garages, and a conventional hinged door would have required wider spacing. This was overcome by the adoption of a sliding cab door for the RT. In addition, a redesign of the pedals eliminated a source of draughts from the slots in which the pedal stems on earlier designs worked. Because of the relative movement between the front-end of the chassis and the body, from which the cab floor was suspended, a good fit was impracticable. On the RT, the pedal stems fitted snugly in holes in a pedal plate which formed part of the chassis and this was surrounded by a sealing rubber which was designed to accommodate the movement in relation to the body.

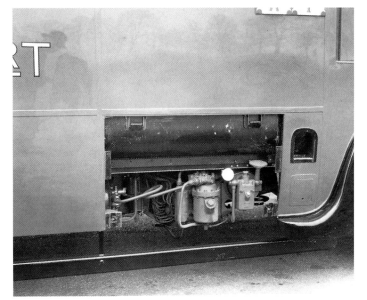

Facing page The RT was slightly heavy-looking at the rear due to the vertical profile being continued up to upper-deck waist level, though the overall effect was strikingly neat and functional. This photograph reveals the staircase window, unique among RT-types, but a feature that had been provided on previous standard classes to allow the conductor to signal when the bus was about to turn right. It was an anachronism, rarely used by then, not fitted on later examples and later deleted from this body, when the offside route number plate was repositioned nearer the rearmost lower-deck saloon window. The graceful curve of the RT's front end is well illustrated in this view. It also shows the sliding cab door, the top of its window curving down to meet the top line of the passenger windows.
LT Museum

Upper left Careful thought was given to the staircase design, with slightly shallower steps to make it easier to climb. This view shows the nearside route stencil, the outline of the characters matching those used on the blinds – consistency of style helped to convey that air of supreme professionalism so characteristic of London Transport at its best. The batteries were housed in the box provided for them under the stairs – the awkwardness of lugging them across the platform led to the provision of external access from the offside on the post-war RT
LT Museum

Left A removable panel on the offside gave access to the filler plug of the oil separator and filler tube of the RP lubricator oil tank. The unloader valve is in its original position on the top cap of the oil separator and access to the oil reservoir is gained via a small floor trap under the front offside seat. The air filter and anti-freeze unit was removed on the production vehicles to the rear nearside corner of the driver's cab.

Right and facing page Official photographs of RT 1 were taken in April 1939 and these show it in the revised livery which was to be special to this vehicle, and lost when it was repainted in the same wartime style as the production vehicles. It was however the basis of the post-war standard style as adopted for new RT-type buses from 1947. At that stage, the seating capacity was 56 but, with an unladen weight found to be 6tons 15cwt 3qr, this had to be reduced to 55 to get the gross weight with full passenger load below the limit of 10½ tons then in force. The set of 164A destination blinds and Sutton (A) garage plates had been fitted to show the complete effect and it is not thought that there had been any intention to operate the bus from there.
LT Museum

The interior of RT 1 set the standard for the whole RT family as built up to 1954, and to remain a familiar sight to Londoners for 40 years. It was to influence many other British buses during that period. The seats were basically the established design as used from 1936, and the pillar finish had also been used on single-deckers from that year, but the excellent forward vision and neat effect of carrying the window sill level across the front bulkhead was new. The ceiling contour was largely inherited from earlier London designs, though more arched than was usual in the lower saloon of other double-deckers of the time. *LT Museum*

The new body was mounted on 27th March 1939, though detail items of completion were still in progress. There was still some debate about the livery, and different styles were tried before arriving at the version finally adopted, which proved to be like the early post-war RT standard. The plan had been for the vehicle to be a 56-seater, like the standard STL, and indeed it was built thus – the series of official photographs taken shortly afterwards show it in this form. Upon weighing, it was found to be marginally above the 10½-ton maximum gross laden weight limit then in force. Accordingly, it was reduced to 55 seats, with 29 on the top deck, and weighed 6tons 15cwt 3qr unladen as recorded on 6th July. The alteration in seating layout, with a single seat on the nearside opposite the top of the stairs coincided with the introduction of a new type of moquette in place of the initial STL-like green. This used square patterns of green, brown, black and red.

The First RTs

Similar general remarks applied to the upper-deck, though here the roof contour was more in line with general practice. The large single window at the rear was something of a departure from the general trend of the time, helping to give an airy interior and also giving a good general view to the rear. *LT Museum*

Bottom left and right The initial seating layout at the rear of the upper deck and the layout after the seating capacity was reduced from 56 to 55 by the installation of a single 'dickey' seat to replace the only free-standing frame on the vehicle. At the time of conversion, the seats were trimmed in a new pattern of moquette (later to be applied to the production batch upon delivery) replacing the STL green moquette with which the prototype had been originally fitted. Although a convex mirror above the staircase was omitted from the design, an amendment to the specification was eventually issued which resulted in the fitting of this item to every member of the class. Note the upholstered top to the side destination box casing, forming an armrest, a feature to remain characteristic of the class.

RT 1 undergoing the statutory tilt test, meeting the requirement that a double-decker must remain stable on a surface tilted to 28° with sandbags to represent the weight of passengers occupying upper-deck seats. The ropes over the body provided for a possible failure but remained slack here, though the vehicle had tilted to about 36° on its springs. The tilting table at Chiswick was much in demand not only for London Transport's own needs but also for the testing of vehicles for several major bus and coach bodybuilders in the London area.

A week after its seating was revised, on 13th July 1939, RT 1 was revealed to the outside world for the first time, when an invited party from the technical and national press, plus representatives of operators and manufacturers, were taken from the Aldwych (near the entrance to Bush House) over a hilly route to Spaniards Inn on Hampstead Heath. The search for sensation was nearly as strong among mass-circulation national papers of those days as today. To their reporters, no doubt RT 1 looked much the same as the typical London buses, largely STLs, they saw every day, though practical points such as the wider windows and the easy operation of those that opened were mentioned in several of the reports. The LPTB had longer-term aims and could afford to be judged by posterity. Even so, no-one could have guessed that RT buses, visually much the same as the prototype, would remain in service until 40 years later. What was remarkable was that even then they seemed no more outdated than typical 10-year-old buses had in 1939; this was the true measure of the almost timeless quality achieved.

RT 1 was relicensed for service on 17th July 1939, being allocated to Chelverton Road garage, Putney, initially for driver familiarisation, and entering service on route 22 from Putney Common to Homerton on 9th August. The decision that this was to be the new standard type of bus had been made and Chelverton Road and nearby Putney Bridge garages were to be the first to receive production vehicles. A new era had begun.

That same day, the cost of constructing RT 1 came up for review by the Works Committee. A sum of £3,750 had been authorised by Special Expenditure Requisition G316, which had been the authority under which the bus was built. It was reported that this figure would be exceeded by £977, thus bringing the total to £4,720. Minute 440 on this item also reported that there would be an underspend on the budget for the experimental account for the year ended 30th June 1939 of about £1,000 and it was agreed that the extra £977 would be charged to the experimental account. The minute did not quote how the £3,750 had been divided between the chassis and body, but the overspend amounted to £787 on the chassis and £190 on the body.

RT 1 was thus a very expensive vehicle by 1939 standards, but the fact that much of it was hand made by skilled craftsmen because so little was to existing production pattern made this unsurprising. That the excess charge was mainly related to the chassis may well be largely due to the additional work done at AEC early in 1939, notably to bring the engine to 9.6-litre form, much as adopted for the production vehicles by then on order.

Two photographs taken at the Aldwych on the day of RT 1's demonstration to members of the press and transport officials, 13th July 1939. After these photographs were taken the party rode on the bus to Hampstead Heath. Austin taxi ELW601 on the right of the upper photo is, like the body of RT 1, still in existence today.

Following the demonstration of RT 1's capabilities to the press on 13th July 1939, two of that day's three London evening papers and four of the following morning's national dailies included brief reports. Most concentrated on the practical aspects, such as the easily-operated winding opening windows, not hitherto seen on a London double-decker – one of the comments acts as a reminder of how prone the previous type of half-drop was to jam in its guides. Most referred to more leg room on the upper deck, suggesting some 'sales talk' by the LPTB representatives, as in reality the differences from the standard STL were no more than marginal – the typical STL14/1 body of 1937-38 had a standard upper-deck seat pitch of 2ft 5in and the figure for the RT2 body type (unlikely to have changed from RT 1) was 2ft 5¼in. The wider windows and the double-skin roof with its better insulation also attracted general attention. The more popular papers favoured 'human interest' stories, such as the presence of a newly-retired driver who had begun his career on horse buses in 1894, not changing to motor buses until 16 years later. More to the point, *The Times* referred to RT 1 climbing the hill to Hampstead "with apparent ease" and referred to the combination of safety, comfort, quietness, long working life, and economic maintenance, indicating quite a far-seeing understanding of the more strategic objectives. *The Daily Telegraph's* motoring correspondent put rather more emphasis on such aspects as better vision for the driver and the light action of the controls. Evidently, he had raised the question of the vehicle's layout, reporting that the "LPTB engineers consider that a rear platform prevents anything but the customary position for the engine and driving axle" – perhaps he was recalling the Q-type, but the wording suggests that he also had in mind the possibility of front-wheel drive. Like the correspondent from *The Times*, the *Telegraph* man referred to the ten route displays, saying "it will be easier to catch one's 'bus when this type is in regular service". The only new displays however were the rear roofbox and the route number stencil under the canopy. *The London Evening Star* was impressed with the actions of the driver during the trial run, referring to him driving what it described as 'the palace on wheels' most of the way with one hand and at times, it seemed, with one finger. Clearly this had showed off the light steering effectively but would LPTB driving instructors have approved?

Mr. Arthur Hill and Mr. Orville Cæsar from America testing the seating in the new bus to-day.

Your New Limousine

Here's the latest luxury saloon by the suppliers of the most popular motors of all . . . the London Passenger Transport Board.

Features of RT1, as it is called, are:— More leg-room on top, wider windows, bigger and better destination signs, oil-driven engine and fluid flywheel, brakes and gears worked by compressed air and rubber mudguards (pedestrians, please note). Picture was taken at a London demonstration yesterday.

America's Bus Chief Tries One of Ours

Evening Standard Reporter

MR. ARTHUR MIDDLETON HILL is president of Greyhound Lines, the enormous road transport concern in the U.S.A., which has a fleet of 2500 motor coaches operating from Atlantic to Pacific, and from the Gulf of Mexico to Canada.

And to-day Mr. Hill, with his vice-president, Mr. Orville S. Cæsar, had a ride in London Transport's latest type of bus.

"Our buses—or coaches as you call them—are naturally a very different proposition from yours," he said to me. "We do long cross-country journeys with perhaps a hundred miles or more between stops. In the open country we travel at 55 to 60 miles an hour.

"But at the same time there are still things we can learn about operating problems, time schedules, and that kind of thing, from the ordinary city bus such as this."

Another passenger on the first run of the new bus was Mr. W. R. Nurse, of Woodstock-road, Forest Gate. He is a retired London bus driver. His 45 years service included 16 with the old horse buses.

MORE LEG ROOM

The new bus, which will go into service immediately, is outwardly little different from existing types. But there is more room inside, the rear platform is bigger, and there is more leg room between the seats.

A double roof has been fitted with the object of making the top deck cooler in summer and preventing condensation in the winter.

Brakes and gear-change are power operated. The driver cannot engage the gears unless there is safe minimum of compressed air in the braking system."

EVENING STANDARD
13th JULY 1939

EVENING NEWS
13th JULY 1939

DAILY MIRROR
14th JULY 1939

London's Latest Bus Has Style

LONDON'S newest bus came on the streets to-day for a trial run, and it had great difficulty in starting.

There was nothing wrong with the engine. It was just kept back by public interest.

Only with the co-operation of the police was the driver able to shake off the messenger boys, bank clerks, postmen and other workers who swarmed round this magnificent vehicle.

WITH U.S. APPROVAL

"Like a trolley-bus," said one messenger boy, forgetful of penny ice brick as he gazed.

"A beautiful job," said Mr. Arthur Middleton Hill. "Just fine," said Mr. Orville S. Ceasar, who was with him.

Messrs. Hill and Cæsar are respectively president and vice of the Greyhound Corporation of the United States. They have been taking a busman's holiday all over the Continent, and they just had to take this trial trip.

THE VOICE OF EXPERIENCE

All present and correct was Bill (otherwise Mr. W. R.) Nurse, a bus driver who started in horses and has just retired from London Transport after 45 years service.

"Would I drive this?" said Bill. "Why, it's dead easy. I don't call it work at all. When I went over from horses to petrol we had to work, I can tell you, and many a night out I've had with a broken-down bus."

His words were justified by the trial trip driver. For most of the way he drove the palace on wheels with one hand—at times, it seemed, with one finger.

ALL POINTS OF VIEW

It may have been that he was the highest thing on the road. It may have been that he could see all round —front, back and skywards. It may have been the compressed air system which makes braking and gear-changing child's play.

Anyway, he was the object of admiration from drivers and conductors all along the route. Some conductors hung from their platforms drinking him and his bus.

From the passenger's point of view this bus has windows which open by turning a handle, improved ventilation, and ten route indication signs. Such a job, in fact, that young London will ride in it for hours, just to absorb it.

NEW TYPE OF BUS FOR LONDON

DEVICES TO INCREASE SAFETY

A new type of double-deck bus which in time will replace those now used by London Transport was given its first public trial yesterday, when with a load of invited passengers it ran from Aldwych to Hampstead Heath and back. Though proceeding along recognized bus routes the vehicle completed the journey in record time without infringing speed regulations. It climbed the hill to Hampstead with apparent ease. In appearance the new bus differs little from those now running, but it embodies every modern device that has been proved to ensure safety, comfort, quietness, long working life, and economic maintenance.

Safety is increased by the braking and gear changing being power-operated by a compressed air system; the driver cannot engage the gears unless there is sufficient air pressure in the braking system. The designers have increased the driver's range of vision in front and on both sides to a maximum by placing his seat higher and further forward and lowering the radiator and bonnet. The driving seat is adjustable, the cab may be ventilated at will, and a new design of floor plate and pedals prevents draughts to the feet

There is more leg room for passengers on the upper deck, and the platform is bigger. A double ceiling will keep the roof cool in summer and prevent the condensation of moisture in winter. The engine is oil driven, the self-changing gear is pre-selected by hand, there is a fluid flywheel, and a complete system of chassis lubrication is operated by compressed air in connexion with the brake pedal.

To enable an intending passenger to tell at a glance whether a bus is the one he is wanting, destination and route indicators on the new vehicle have been increased in number and their design improved. No fewer than 10 signs are exhibited—five giving the number of the route, three details of the route, and two the destination. The windows of the bus are wider than those in present buses and easily adjustable by winders. The mudguards are made of rubber.

NEW 'BUSES FOR LONDON

GREATER COMFORT AND UTILITY

By Our MOTORING CORRESPONDENT

I was a passenger yesterday in the latest London 'bus, the first of a new type, known as R.T.1, which, if tests in service in September prove satisfactory, will set the standard for London 'buses for several years to come.

It may set a standard for America, too; for among the passengers were Mr. A. Middleton Hill and Mr. O. S. Caesar, president and vice-president of the largest 'bus and coach corporation in the United States.

An interesting innovation in the new vehicle is its rubber mudguards. The general design is orthodox, for the L.P.T.B. engineers consider that a rear platform prevents anything but the customary position for engine and driving axle.

Passengers will have more leg-room upstairs, more space on the platform, wider windows, and better route indicators. It will be easier to recognise one's 'bus when this type is in regular use.

There are 10 signs for passengers' guidance; five giving the number of the route, three giving details of the route traversed, and two showing the destination.

The driver's range of vision has been increased by placing the driving seat higher and further forward and lowering the radiator and bonnet.

Braking and gear-changing are power operated by compressed air, arranged that the driver cannot engage the gears unless there is sufficient air pressure for the brakes. Slight physical effort is required in braking, and all pedals are just "heavy" enough to ensure accurate control.

Other factors to ease the driver's task are an adjustable seat, adjustable ventilation, no draughts round his feet and insulation from engine-heat.

NEW TYPE 'BUS.—One of the L.P.T.B.'s new omnibuses photographed while on view in Aldwych, London, yesterday.

London's New Bus

A new London Transport double-decker bus made a demonstration trip to-day from Aldwych to Hampstead carrying, among others, the president and vice-president of the American Greyhound Corporation, which is the big American coast-to-coast motor-coach company, and also a retired London Transport driver of forty-four years' experience with horse and motor buses. The American passengers admired the bus but would not admit that this new, smooth vehicle surpassed anything that New York or Chicago could show.

It is the first of the new type with which the London Transport Board is going to supersede the existing types. Like its predecessors, it carries fifty-five passengers, but the upper-deck, at the insistence, one is told, of Lord Ashfield, gives a good deal more leg room. And from the passenger's point of view the other improvements are more indicators of route and destination and bigger windows which wind up and down at the turn of a handle, so the sight of perspiring men struggling with both hands to press open a London bus window will eventually become a spectacle of the past.

For the driver the new advantages are a higher driving seat, giving better vision, and a safer braking system operated by compressed air. The impression, though a misleading one, is of a longer vehicle—something more like the lines of the present trolley buses. The wings are all of rubber—not, however, to preserve obstructing pedestrians and vehicles; but simply because rubber is less vulnerable than metal. This bus, so the veteran driver assured us, is easy to drive—easier, he said, than a horse bus.

THE TIMES
14th JULY 1939

DAILY TELEGRAPH
14th JULY 1939

MANCHESTER GUARDIAN
14th JULY 1939

CHAPTER TWO

Putting the RT into Production

The decision to go ahead with the new type of bus was made even before any significant operational experience of the prototype in its initial ST 1140 form had been obtained. That vehicle did not receive its temporary open-staircase body until 1st July 1938 and did not enter service until the 13th, yet authority was given by the LPTB to order the initial production batch of 150 chassis for delivery in 1939-40 that same month.

Admittedly, the body order did not come until November 1938, when a further 188 chassis were ordered, and 341 bodies were ordered for the combined total of 338 chassis, following the usual practice up to that period of providing a float of spare bodies to allow for the longer time required for body overhaul in relation to that needed for chassis. The float specified was thus a little under 1 per cent rather than the traditional 3 per cent that had been in force previously. Even at that date, the prototype RT chassis was still running with its temporary body and the production body order was thus based on no more than drawings, a model and whatever may have been produced in mock-up form at that stage.

London Transport was apt to be criticised for its extravagant-seeming specifications, but the initial estimates made when the approval for purchase was given by the Board on 3rd November 1938 were that the combined total of 338 RT buses would cost £950 each for the chassis and £795 for the body, with three spare bodies at the same price. The same requisition, G325, included 17 STL-type buses (part of the 132 of the 15STL16 type built in 1939) and at that same date they were to cost £850 for the chassis and £650 for the body apiece. At that date AEC was quoting no less than £1,425 for a standard 7.7-litre Regent chassis with crash gearbox and vacuum brakes in published price lists, though it is clear that AEC 'built in' a substantial sum to allow quite heavy discounting. Even allowing for the fact that the specification was not then finalised and some subsequent alterations were to increase the actual cost, the LPTB thus planned to pay less for buses of the most advanced mechanical design in service anywhere in Britain than was being quoted for buses of quite basic design.

Considerable sums had been spent on the development of the RT, the figure for AEC's share of the work alone being quoted as about £30,000 (about £1million in today's money) but the intention was that London Transport would repay this cost and its own investment in the project by ordering at least 1,000 vehicles.

In pre-war days, AEC's method of dealing with substantial orders for non-standard chassis, or indeed other work where an extensive list of parts to be added and deleted was required, was to issue what was called an XU list. These were themselves given serial numbers, and XU1631 with 'Application' quoted as 'LPTB' and the description 'Parts required for 1939 RT Regent' was issued on 14th October 1938. Ever since his days with AEC in 1951-5 and until very recently, the author was under the impression that this number related to RT 1. This was because it appeared in the space headed 'Originated for' which formed part of the standard AEC drawing sheet in cases where the item shown had started life as a part used in the original RT design.

In fact XU1631 related to the first batch of production chassis, in other words RT 2-151. Most of the same drawings had indeed been first used for RT 1's chassis, the drawing date often going back to 1937, but they would have been issued for that purpose as parts for an experimental project, and XU1631 would have been inserted in place of the original experimental order number as part of the process called 'productionising' when drawings were issued for volume manufacture.

The date of issue of XU1631 itself gives the clue to this, since by then the first chassis was already on the road in its temporary ST 1140 guise. What is particularly interesting is that a further number, XU1651, was issued, with the application quoted as 'O661RT' and the description 'Parts for 1939 chassis for LPTB, 188 sets, programme UB' and the date 25th May 1939. Clearly this related to the second order that was destined not to be fulfilled because of the war, but its existence as a separate parts list suggests that some significant differences in design were planned.

From about the spring of 1938, there was an increasing element of uncertainty in the future plans of businesses of many kinds in Britain. The threat of war with Germany, as its Nazi government ordered the annexation of neighbouring territories, had seemed worryingly close and even though the imminent danger appeared to recede for a time, the programme of re-armament was given higher priority. AEC, like most of the leading makers of commercial vehicles, was asked to devote more of its design and production capabilities to military vehicles or engines, with consequent delay to other work. It was fortunate that the RT project had got through its prototype stage before this factor became too serious, and production of the 150 chassis began not far behind schedule in the summer of 1939.

The chassis design was slightly altered from that of RT 1. Most obvious was a reversion to a frame design with the side-members continued behind the rear axle to support the rear of the body, this being a consequence of LPTB's decision not to proceed at that stage with a production version of the largely metal-framed body design developed for RT 1. The standard form of construction then being used for body production at Chiswick works retained the traditional type of wooden framing, using ash reinforced by metal flitch plates where considered desirable. It had been shown to be capable of giving about a ten-year life, given the then normal programme of intermediate overhauls, and that was considered adequate in relation to the expected life-span of a bus.

It is self-evident that the LPTB was aware of the benefits in terms of reduced maintenance of a well-designed metal-framed body from the expenditure of a considerable sum on the body design for RT 1, but the body shop at Chiswick was geared to the production of timber-framed bodies and it was decided not to disturb this at that time. The difficulty in getting supplies of suitable sheet steel, mentioned in relation to the 10T10 bodies, may still have been a factor. The risk of delay in completion of the vehicles, due to the time required to re-organise the shop for metal body construction, and the general atmosphere of uncertainty in late-1930s Britain may also have had their influence. As it turned out, the change in production facilities was never made, since even at the end of the war the repair and overhaul backlog in a fleet largely beyond its planned life and subject to neglect due to war circumstances was so great as to cause London Transport to give up any idea of returning to bodybuilding.

The decision having been made, the choice of wood-framed construction to the 'improved' Chiswick specification produced a lighter vehicle, the unladen weight with 56-seat capacity (in the standard arrangement with seats for 30 passengers upstairs and 26 down) coming out at 6tons 12cwt (exactly the same as the standard STL buses of 1936-38) which allowed the vehicles to be well within the gross weight of 10½ tons. Although the timber-framed body was less durable, this must still be counted as quite an achievement, bearing in mind the specification and finish of both chassis and body.

Chassis O6616750, destined to become RT 2, is seen here with temporary wooden dash panel and battery box, almost ready for its delivery run from AEC's Southall works on 16th October 1939. In this case, the various units and individual parts were in 'as manufactured' finish, no overall painting of the chassis had been carried out, and the front wheel-nut guard rings had yet to be fitted. The war had begun, and the corner of the kerb in the back-ground had received white stripes to make it more easily visible in the blackout. The chassis design had been altered in quite a number of minor details from that of RT 1, as can be seen by comparing this view with that on page 19. *AEC*

The opportunity was taken to tidy up the body design in quite a number of minor details, though the strikingly modern overall impression of the prototype was maintained. The reverse curves at the rear of the mudguards were eliminated except for the offside rear, simplifying the design and producing an outline more in keeping with contemporary trends in car mudguard fashion. On the standard version, the profile was made even smoother than on RT 1 by eliminating the very slight projection of the front roof dome over the upper-deck front windows. The driver's windscreen frame was given rounded outer corners at the top, a typical example of the attention to detail given to matters of appearance. On the other hand, the deepening of the valance to the nearside canopy was perhaps a slightly retrograde step. The squaring up of the driver's cab door window outline, which had matched the curve at the rear corner of the canopy on the nearside on RT 1, may have been dictated by the need to avoid obstructing a tall driver's vision when looking to the right at a road junction. The elimination of the push-out ventilator in the panel above the windscreen seemed a justifiable simplification in view of the hinging of both upper and lower screen glasses to give generous ventilation if needed in summer.

The method of construction allowed certain variations to be tried out and the bodies fitted to RT 25 and RT 100 later became the subject of Experiment S5489 which reinstated the slight peak effect of the prototype at the front of the upper deck. Rubber mudguards, which resisted damage in minor accidents, were in favour to some degree within the bus industry at the time and were specified for the 2RT2s, steel being quoted as an acceptable alternative in the event of a shortage of rubber. At first no offside headlamp was provided, and although this proved logical in the early days of the war when only one masked headlamp was permitted, the fact that a general arrangement drawing of the completed design dated July 1939 shows a headlamp only on the nearside indicates that it had been a peacetime intention, despite RT 1 having been built with the normal pair.

The signalling window on the right of the platform, another rather anachronistic London feature originally provided to allow the conductor to signal a right turn, was also omitted. On the other hand, these buses still used the so-called bulb horn, regarded as obsolete elsewhere by the early 1930s, and seeming completely out of character on these ultra-modern vehicles. Only London buses and taxis still used so old-fashioned a means of audible warning, this being a Metropolitan Police requirement. Other simplifications were the deletion of the third opening window on the nearside upstairs and of the ventilator on the nearside of the roof.

On RT 1, the ventilating louvres provided internally at the front of both saloons had means of closing, using the conductor's budget key, but for the production vehicles this facility was provided only on the lower deck. This permanent ventilation may have been intended to alleviate the accumula-

tion of tobacco smoke so typical of the upper deck of many buses in those days, but, on some buses at least, newspaper was soon stuffed into the ventilator. RT 113, for example, had it permanently blocked with a copy of the Evening Standard dated 8th May 1940, one week after it entered service.

Fresh thought was given to the livery, though the end result was a rather odd mixture which seemed to suit the RT body rather less well than that adopted for RT 1. A reversion was planned to the silver-grey roof colour as standard on earlier types of London bus, though the rear dome and front route box surround were to remain red and there was a fine red line separating the grey from the off-white of the upper-deck window surrounds. There was logic behind the change, in that silver was less absorbent of summer heat, but while well suited to the STL, it seemed less satisfactory for the RT.

As it turned out it was decided to adopt a dull grey shade for the roof panels to make buses less conspicuous from the air under wartime conditions. Another change was the use of the off-white for the lower as well as upper deck window surrounds. Within, the general appearance of RT 1 had been judged very effective and was generally unchanged, the new type of seat moquette adopted when that vehicle had been modified before entering service now being standard.

A significant change in the chassis specification was to replace the Bosch compressor and other equipment for the air-pressure brake system with components of Clayton Dewandre make, though AEC chassis build sheets show the compressor as 'Clayton Bosch RC7', and the valve operating the low-pressure 'red flag' unit in the cab was of Westinghouse make. This was no doubt seen as both normal choice of British-made equipment where available and also a sensible precaution in the event of war. Clayton Dewandre had been more especially associated with vacuum brakes, the servo of which was based on a Dewandre patent, but the firm was widening its range of products to cover both types of equipment.

As it turned out, this switch in supplier proved essential; for before the first production chassis had been completed, war had been declared on 3rd September 1939. The first two chassis to be delivered from AEC, to become RTs 15 and 23, arrived at Chiswick on 13th October. Thus what are often called the 'pre-war' RTs belong to the early wartime period. Body construction was also in hand and several bodies were completed before the end of the year but no vehicles entered service before January 1940.

The common impression that buses built under wartime conditions were necessarily sub-standard in specification or finish is quite erroneous in regard to those completed before 1941, and never more so than with this splendid class of bus. Most of the parts needed to complete them were in existence or on order at the outbreak of war, and at first disruption of civilian work was often more in the nature of delay than cancellation, especially since effective public transport was of obvious importance.

The First RTs

AEC chassis numbers were issued on confirmation of an order, and thus frequently do not represent the order of construction. At that period the RT-type Regent was included in the same series as standard models, with O661 prefix (the O signifying 'oil engine'). Thus RT 1 was O6616749, numerically coming immediately after the STL 2516-2647 batch of 7.7-litre Regents for London Transport, themselves a cut above the standard STL with their flexible engine mountings, which were O6616617-6748, entering service during 1939. The RT 2-151 batch were O6616750-6899, following on immediately after RT 1, though actually separated by over 16 months in date of chassis delivery, during which many other Regents for other operators were built, having numbers both before and after the London batches.

London Transport's coding system for its vehicle types conveyed chassis variations by a prefix number to the type letters and body variations by a suffix number. Thus RT 1 was chassis type 1RT and body type RT1, the complete vehicle being described by the overall code 1RT1. The 150 production buses were 2RT2, both chassis and body changes being signified by a new code. While dealing with numbers, it is convenient to explain other aspects of London Transport methods. The Board had its own series of body numbers, the body built for RT 1 being 18246 in an unclassified series that went back to early LGOC days; it fell between the bodies for the two batches of 10T10 Green Line coaches, though these were all in service before RT 1 was completed. It was decided to start a fresh series in 1939, beginning with the TF-class single-deckers, followed by the STL-class buses mentioned above, the CR rear-engined Cub 20-seaters plus various acquired buses. The RT2-type bodies had numbers 280 to 429.

LPTB practice at that time was to keep the fleet (bonnet), registration and chassis numbers in sequence within batches and thus RT 2-151 were registered FXT177-326 and had the chassis numbers quoted above, all in tidy order throughout the batch. However, the vehicles were not completed in numerical order and the body numbers were not in sequence with the other numbers. It appears that body construction had reached a point where quite a stock of bodies was ready for mounting before that process began, and the earliest vehicles completed took bodies with a range of numbers.

Chassis delivery had continued at about five per week and by the end of 1939 all but nine of the chassis for RT 2 to RT 62 had arrived. The actual order of delivery was quite erratic, especially in the early weeks, and after the chassis for RTs 15 and 23 arrived on 13th October, that for RT 2 followed on the 16th. Up to a point, this was not surprising, especially with vehicles of a type new to the production line. One or two of the early chassis may have been held back for development and testing by AEC's experimental department, especially as the brake system was not the same as on RT 1. It is significant that RT 5's chassis did not arrive at Chiswick until 4th December, yet was among the earliest to be bodied, even if not quite the first.

However, it is likely that the chassis came down the line in numerical order or nearly so, in AEC's usual fashion (normally the only exceptions to this were where there was some variation in specification), and the erratic arrivals at Chiswick suggest that there might have been problems of one kind or another in quite a number of cases. 'The war' tended to be a convenient excuse for all kinds of delay and clearly, even in late 1939 when there was very little action, military needs would have had priority. Delays in supply of specialised parts may also have held some chassis back temporarily. However, the strange order of delivery does raise the possibility that AEC's own engineers may have been having difficulty in getting some of the chassis to perform properly.

There would have had to be a process of learning about such a model, with its many new features, and in particular the whole air system and its foibles, to be undergone by both AEC and LPTB staff. The issue of a Maintenance Bulletin for the 2RT2 by the London Transport engineering department was an indication of official recognition of this need. It ran to 40 pages of general information, with additional sections on lubrication (8 pages) and the air pressure system (18 pages). It is a revealing document in terms of the philosophy then in force as well as the detailed information included and is the subject of an appendix in this volume.

With hindsight, it seems as if the fact that these 150 vehicles were to be the only production examples for the time being may have been a blessing in disguise, giving time to sort out the problems experienced before the huge post-war RT programme got under way. The balance of 188 ordered in November 1938 was not put in hand and indeed, even after the war, 'disappeared' amid the post-war deliveries, none of which matched this quantity, though there are indications that AEC may have planned to build such a number as its first post-war batch for London Transport until circumstances caused it to rearrange its orders.

Official records give the impression that the Certifying Officer at Chiswick had a busy day on Monday 1st January 1940 when fifteen 2RT2s received their PSV Licences of which six, RTs 15, 16, 19, 39, 43 and 44, also gained their Certificates of Fitness. In the days which followed the nine remaining vehicles also received their CoFs; RTs 5, 25, 37, 38, 47 and 56 on 2nd January; RTs 30 and 39 on 3rd January and RT 20 on 5th January by which date all had been licensed for service from Chelverton Road, Putney (code AF) where RT 1 was already operating. The fifteen new buses were sufficient for the complete allocation on route 37 (fourteen plus one spare) and former members of Chelverton Road staff recall that this was the route on which they were almost exclusively used until further deliveries in March. Photographic evidence exists however of RT 5 working at least once on route 22 during this period, to which RT 1 was officially allocated. Body numbers for this initial batch of vehicles were allocated numerically as each was completed and thus RTs 15, 19, 16 and 20, which were the first of the class taken into stock on 28th December 1939, received the four lowest numbers (280-283). However, RT 2 with body number 317 was not taken into stock until 29th January and remained unlicensed at Hanwell garage until 20th March when it was transferred to Chelverton Road and licensed for service the following day. This was not an uncommon pattern for some of the earlier vehicles and would have been convenient for attention by AEC fitters from nearby Southall works, who could have driven the vehicle, if need be, on trade plates. Among other early numbers, the chassis of RT 11 had been delivered on 3rd November 1939 but was not officially taken into stock as a complete bus until 13th March 1940, eventually entering service on 1st May.

The 2RT2 buses began to enter service in January 1940. This view at Putney Common of RT 5, which was licensed for service on the 5th of that month for operation from Chelverton Road, conveys the look of the type when newly in service. Prior to receiving RTs, Chelverton Road had been 100% STL. Full destination displays were still being fitted, in this case for route 22, which apart from RT 1 did not have a scheduled allocation of RTs. The white paint on the mudguard soon showed minor blemishes. *Lens of Sutton*

RT 35 at East Dulwich Grove on 31st March 1940, just ten days after it was licensed for service at Chelverton Road garage. Former staff there recall that the first deliveries of 2RT2s were used mostly on the 37, but this vehicle was one of those delivered to the garage to permit the conversion from STL of routes 28 and 30. After experience in service with the first fifteen 2RT2s, a small spotlight was fitted below the offside dumb-iron. Officially described as a 'P-type Pass Lamp', its output was restricted to meet blackout regulations. Later in the war, a conventional offside head-lamp was fitted, as by then two masked headlamps were permitted. RT 35 was photographed during the 'phoney war' period before air attacks on Britain had started and only one of the rooms in the houses and flats in the back-ground has its windows taped as protection from blast.
John Bonell

Among the 'first day' contingent, RT 19 was taken out of service after only 3½ weeks at Chelverton Road and returned to Chiswick on 27th January 1940, before going back to AEC, at that firm's request, for modification and a 2½-year career as a demonstrator. This was an unusual course of events for a London bus, no doubt caused by war conditions when the construction of a demonstrator owned by AEC may have been impractical. It made an extensive series of visits to operators in almost all parts of Britain in that time, clearly indicating AEC's interest in the model as a basis for future business with operators outside London. The point was emphasised by the delivery to Glasgow Corporation in February 1940 of the one RT-type chassis (O6616963) to be built for other than London Transport during that period, fleet number 723, registered DGB371. This, which had Weymann bodywork to a modified version of Glasgow's standard of the time, had been intended for display at the Commercial Motor Show planned to be held at London's Earls Court in November 1939 but cancelled due to the war. More information on both episodes appears in the next chapter.

No further members of the class were delivered to Chelverton Road until the third week of March, vehicles from the initial batch of fifteen having been used to displace some of the garage's allocation of STLs. Route 22 was never officially converted to 2RT2 operation and became the responsibility of the new garage in Gillingham Street, Victoria, which opened on 20th March. The next day a further forty 2RT2s were licensed for service from Chelverton Road which, together with

The war had not made too obvious an impact on this scene showing one of Chelverton Road's allocation of 2RT2s at Clapham on route 37 in May 1940. Not many intending passengers are in uniform, and gas masks are mostly conspicuous by their absence, despite official exhortations that they should be carried. Forming an orderly queue, which was soon to become accepted behaviour under wartime conditions, had yet to catch on. The full destination display was being used at that early stage, including the rear roof box, and this view shows that it was too high to be much use for anyone wishing to check the route number just as they were about to board. Route 37 was the only route to retain an allocation of 2RT2s throughout the whole of their existence in Central Area service. *LT Museum*

the prototype, were employed to convert route 30 (20 vehicles) and partially convert route 28 (19 vehicles), allowing two spares.

Deliveries to Chelverton Road garage continued during April (five buses) and May (13 buses), these being used to displace the remaining STLs then allocated to routes 28 and 72. Official figures provided for 15th May show that, in addition to the number of 2RT2s used to convert route 72 (10 vehicles), a decrease in the number of workings on route 28 (15 vehicles) and slight increases in those for routes 30 (21 vehicles) and 37 (15 vehicles) made a total of 61 2RT2s being allocated from the total of 73 already consigned to the garage.

The introduction of 2RT2s on route 14 began in May, when eighteen of the type were delivered to Putney Bridge (code F)

and licensed for service from the first of the month. This quantity represented just a little over half of the garage's vehicle allocation to this route and, in order to increase the number, it can safely be assumed that the first of many movements between the two Putney establishments took place. As a result, eight 2RT2s (RTs 27, 40, 51, 55, 60, 73, 79 and 102) transferred to Putney Bridge from Chelverton Road bringing the official figure for route 14 on 15th May 1940 to 26. A further batch of seventeen vehicles was licensed for service during June and used to increase the route 14 workings to its maximum of 35, the deposed STLs mainly moving to Harrow Weald and Willesden garages. As a rule of thumb, the 2RT2s which had entered service at Chelverton Road garage carried the earlier fleet numbers.

The vehicles were generally well-liked by staff and passengers, but the braking systems were behaving erratically, inevitably a source of worry. The problem centred on the rotary Clayton Dewandre compressor, which was proving unreliable. The principle of a rotary air pump was not new, since the exhauster provided on nearly all oil-engined buses with vacuum brakes was effectively a compressor of this type acting in reverse. Moreover, such units gave little trouble, but Gavin Martin, bus engineer and author, has pointed out an aspect of the design which may have aggravated the problems. The mounting of the dynamo under the saloon floor and to the rear of the engine was an idea which Chiswick tended to favour and adopted for the RT; it was driven from the timing case via a universal-jointed shaft. In those days long before alternators had been developed for road-vehicle use, it provided a means of finding room for a large enough dynamo for the heavy electric lighting load of a city bus, often running at low speed and with frequent stops and starts.

The drive was at fairly high speed appropriate to a dynamo, so by putting the compressor on the front of the dynamo, this too was driven at dynamo speed, considerably faster than an engine-mounted exhauster, driven at half engine speed by the injection pump drive. In one sense, this may have been thought beneficial, helping to increase the compressor's output to meet the quite heavy demand for air to operate both brakes and the gear change in city traffic. However the high running speed and inadequate output caused overheating, leading to trouble with bearings and oil seals. There was a problem with oil carryover into the air system and this led to blocking of the unloader valve, adding to the trouble.

The result was erratic brake behaviour and, understandably, the initial reaction was to modify the design of the compressor to incorporate uprated bearings and oil seals. Kits of parts were put on order, but by this stage in the war it was becoming difficult to obtain delivery for such material within a reasonable time, and conversions proceeded slowly.

This photograph was another of those taken by AEC of the chassis of RT 2 in October 1939, shortly before delivery from Southall to Chiswick. This view shows the original location of the dynamo, alongside the propellor shaft, with the compressor (recognisable by its ribbed casing) in front of it, both being driven by a smaller propellor shaft from the engine timing case. This original form of compressor proved to be inadequate and much of the trouble that plagued the type in its early years centred on this unit. Also visible at the rear of the engine is the fluid flywheel, largely shrouded by the rear engine mounting – the fluid flywheel gland was another item which needed quite frequent attention under London operating conditions – an improved design appeared in the 1950s, but this was too late for the 2RT chassis. To take this photograph, the temporary wooden 'structure' provided for testing and delivery, as visible in the picture on page 37, had been removed and the fuel header tank and filter leant against the front nearside of the engine.

The chassis had all arrived from AEC by May 1940. The last five, RTs 147-151 (chassis numbers O6616895-99), had experimental gearboxes with a smaller drum diameter of 7.75in instead of the standard 9.1in (AEC list XU1677 of March 1940, and London Transport experiment S5335). This was an all-new design produced in Coventry in 1939 by Self-Changing Gears Ltd, the company set up by Walter Gordon Wilson, the inventor of the Wilson gearbox. The SCG works built prototypes and small-quantity orders for gearboxes of Wilson design primarily to promote the sales of licences for larger-scale manufacture to companies like AEC who wished to use them in their products.

The internal design of an epicyclic gearbox is quite tightly constrained, but the reduced diameter of the drum assembly, and hence the whole gearbox, allowed weight to be reduced and in those days that was critical, as evidenced by the need to reduce RT 1 to 55 seats to bring it within the limit on gross weight before it could enter service. To keep the tooth stresses on the smaller gears within acceptable limits, the first gear train had five planet gears instead of the three used on previous designs. This was to be a feature of later Wilson-type gearboxes designed by A.A. Miller and produced by AEC and Leyland from the mid-1950s. Another innovation, apparently also related to the reduced diameter, was the adoption of dry-sump lubrication in which the running gear did not dip in the oil as on the standard version. A separate oil tank would have been provided, with an additional pump to circulate the oil.

Originally six 7.75-in drum gearboxes were to be built by Self-Changing Gears in Coventry, the initial plan having been for these gearboxes to be tried in STL buses, three being designed for operation by hydraulic means and three by compressed air (the latter presumably to have been used in the STL buses running with experimental compressed air brake systems). But the hydraulic version was found to need excessive pressure and it was decided to switch the experiment to 2RT buses. The boxes had been part-completed when war broke out and SCG, overloaded with urgent military work, passed them to Chiswick for completion. This was done, though not without some difficulties with materials and the satisfactory fit of various parts, and some of the failures that were experienced in service were attributed to this.

Five of the 7.75-in gearboxes were completed, the sixth set of parts probably being kept as spares, and these buses entered service among various other 'late starters' from March 1941 (RT 147) to February 1942 (RT 151). So far as is known they retained these gearboxes until about 1946, when standard D140 units were fitted – that in RT 148 is recorded as having been changed in November. The report on experiment S5335, written in June 1946, stated that the 7.75in gearboxes ran hotter than the standard D140, but an advantage was that they were quieter. The dry sump lubrication was not considered to have produced any advantage. However, the smaller-diameter running gear had withstood the loads imposed and was considered worthy of consideration for the future. In fact, it seems never to have been pursued, the pressure for weight reduction having eased amid a general acceptance of higher weights, originally arising from war circumstances.

Approval in principle for another gearbox experiment which would have concerned the 2RT buses had been agreed by the Works Committee on the basis of a proposal by Durrant at a meeting on 22nd May 1939, but evidently did not go ahead at all at that stage. Following earlier work on a bus with 'a semi-automatic type of gear-changing equipment' (STL 760) using a design patented by Self-Changing Gears, drawings had been prepared in collaboration with that concern for a production version. It was proposed that twelve RT buses be so equipped at a cost of about £1,200, though Durrant was to submit a further recommendation to ensure that the LPTB was adequately recompensed for its experimental work if the equipment was to be marketed on a commercial basis. Even at that date SCG was already heavily committed on military work, and no doubt this experiment fell by the wayside. It may have been intended for trial on some of the 188 buses never built. After the war further experiments did take place with what was sometimes called semi-automatic gearbox control – more accurately described as 'direct selection' – with no 'clutch' pedal and engagement of each ratio in direct response to movement of the lever. RT buses were used, but by then of the post-war 3RT type, the first being RT 778.

Reverting to the situation in 1940, body mounting and entry into service lagged somewhat behind chassis delivery. The supply of new RTs to garages ceased entirely for the time being from mid-June, at which time 43 buses out of the batch of 150 were yet to enter service, though the highest fleet number to have entered service was RT 139. It seems very probable there was unease at putting more buses into service until the brake problems were overcome, although other delays due to war conditions may have also been a factor.

The brake situation became such that London Transport was unwilling to risk the danger of an accident, and after delicensing 20 vehicles of the class on 1st July 1940, had withdrawn a further 68 from service by 1st August. The twenty vehicles which remained were RTs 4, 7, 9, 10, 11, 13, 23, 24, 35, 44, 48, 59, 68, 85, 94, 103, 108, 110, 125 and 139, of which twelve had received the necessary modifications and the other eight were completed soon after. These were soon joined by RTs 39, 117 and 137 which, although having been withdrawn on 1st August 1940, were returned to service after only a few weeks in store; RT 39 on 1st September, then RT 117 and RT 137 on 19th September. Two days later, RT 69 and RT 100 were licensed for the first time bringing the total number of vehicles available to 25, all of which were allocated to Chelverton Road garage for service on route 37 (15 vehicles) and route 72 (10 vehicles). However, on 3rd December RTs 4, 10, 39, 48 and 68 were transferred to Putney Bridge, presumably to help address worsening vehicle shortages.

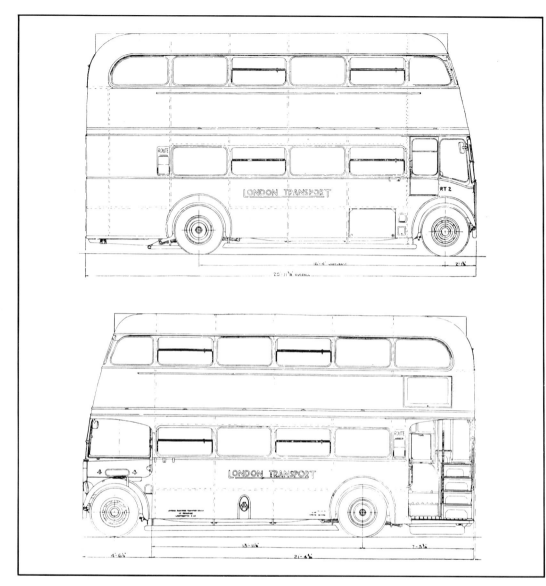

These general arrangement drawings, based on LPTB originals, convey many details of the appearance of the completed 2RT2 vehicles – the beading over panel joints is indicated by dotted lines. *Model Railway Constructor,* realising the importance of the type, chose RT 36 for its front cover in July 1940 and intended to reproduce drawings similar to those above in its August issue, but, in the event, readers had to wait until the June 1956 issue for the publisher's promise to be fulfilled.

At the end of 1940 the vehicle position could be summarised as follows:

Prototype RT 1 (in store)	1
Vehicle on loan to AEC (RT 19)	1
Vehicles available for service	25
In store (following withdrawal)	84
In store (completed but yet to enter service)	28
At Chiswick (RTs 89, 120, 134, 142, 143, 145, 146)	7
Vehicles still to be completed (RT 147-151)	5

The two main storage areas for the withdrawn 2RT2s were situated in the former AEC Works in Forest Road, Walthamstow, and 80 Clapham Road SW9, where the recently vacated London Terminal Coach Station was situated; 45 of

THE MODEL RAILWAY *Constructor*

Vol. 7.   JULY, 1940.   **No. 76.**

A London Passenger Transport Board " R.T." Type Bus.
(4 mm. drawings for this will appear in the August issue.)

PRICE SIXPENCE.

the class together with the prototype were stored at the latter establishment which appears to have been used as a satellite of Chiswick Works, 31 undergoing air compressor modifications here before returning to revenue service. It is interesting to record that three of these virtually new and advanced buses were pressed into use as mess rooms for the newly established Home Guard. Younger readers who associate that organisation with the 'Dad's Army' TV series should bear in mind that in real life, the Home Guards' duties, once basic training had been done, were largely a matter of waiting about in case of a warning of enemy invasion. The RT buses' service duties were taken over by ST-type petrol-engined AEC Regent buses dating from 1930-31, which had been stored due to early wartime cuts in services.

The small number of RTs in service at the end of 1940 bears witness to the difficulty of getting the new parts. By that time the war situation had worsened considerably, France having fallen to German attack and Britain facing the threat of invasion virtually alone. Understandably, getting non-standard civilian vehicle parts made was far from easy.

The temporarily laid-up buses gradually returned to service as the modifications were carried out, but delivery to garages of further new buses did not resume until March 1941. Five entered service that month, followed by a dozen in April, but the work of completing the remainder was still subject to delays, largely related to completion of the bodywork.

Whatever the intended wartime function of the London Terminal Coach Station, its existence as temporary accommodation for 2RT2s came to an abrupt halt on 11th April 1941, following the departure of the last fifteen 2RT2s from this store, thirteen transferring to Chiswick Works for modification and two directly to Putney Bridge garage for service. The prototype was released on 11th April bound for alterations at Chiswick Works before seeing further service at Chelverton Road and Victoria garages.

The 2RT2s stored at 80 Clapham Road were the first withdrawn vehicles to be modified and by early April 1941 Putney Bridge was able to field 22 2RT2 buses on route 14, whilst Chelverton Road's allocation for route 37 had risen to 19. Obviously the vehicles moved from the London Terminal Coach Station had been put to immediate use and a further influx of former inmates from this establishment via Chiswick Works, over the next few weeks, increased the number available for service on route 14 by ten and route 72 by five. A complete list of the vehicles stored at the coach station and other establishments around the capital is included in the appendices.

The modifying of air compressors in preparation for the 2RT2s' return to service continued. Chiswick Works took a selection of vehicles from Forest Road, Walthamstow and the four LT garages used for long-term storage: Potters Bar, Staines, Windsor and Reigate. Nunhead and Merton garages also became involved in the modifications as did Putney Bridge and Chelverton Road garages to which all the withdrawn members of the class had originally been allocated and where the majority eventually returned.

Three further new RTs were delivered in May 1941 and one in June, by which month Putney Bridge's small allocation on route 85 was converted to RT for the first time. There was then another pause until September, when seven new buses took to the road, followed by ten in October. The October 1941 allocation book shows RTs allocated to routes 14 (33 buses), 74 (11 buses, 85 (six buses), 93 (ten buses) and 96 (ten buses) from Putney Bridge garage, routes 30 (21 buses) and 37 (19 buses) from Chelverton Road and the addition of Victoria garage, which had a scheduled requirement of 18 RTs for route 52. Apart from the 52, RTs were also new to routes 74, 93 and 96, though the 74 had had them at weekends since June.

NUMBER OF VEHICLES AVAILABLE FOR SERVICE AUGUST 1939 – OCTOBER 1942

Month	Into Service	Delicensed	Relicensed	Total
8.39	1			1
1.40	15			16
2.40		1		15
3.40	40			55
4.40	5			60
5.40	31			91
6.40	17			108
7.40		20		88
8.40		68		20
9.40	2		3	25
3.41	5		10	40
4.41	12		19	71
5.41	3		2	76
6.41	1		3	80
9.41	7		22	109
10.41	9		25	143
11.41			3	146
12.41	2		1	149
2.42	1			150
10.42			1	151

An alteration advice to fit P-type pass lamps to the whole class was issued by the Drawing Office at Chiswick on 7th March 1940; the initial batch of fifteen vehicles having entered service without this interim modification. However, it was not until early October 1941 that conventionally placed offside headlamps began to be fitted and it would appear that RT 52 is about to have its lighting improved, albeit only slightly as a headlamp mask will immediately be installed on the new fitting due to the prevailing blackout restrictions. Later in the war RT 52 was withdrawn from service as a result of enemy action, its chassis spending some time with AEC. *Alan B. Cross.*

RT 1 in Central Hill, Upper Norwood on Good Friday 1943. The bus had been overhauled at the beginning of the year, emerging early in February in the standard 2RT2 livery. Most of the vehicle's short life span of six years and eight months was spent at Victoria (Gillingham Street) garage, where it was based from October 1941, apart from short-term loan to Putney Bridge garage towards the end of the war. When this photograph was taken, RTs were not officially allocated to the 137.
Bennet Jordan/Peter Hoskins

Below right An unidentified 2RT2 provides an interesting background for this view of a new uniform style for 'clippies' taken during the early stages of the bombing. Of note is the window netting, as originally applied, with both sections of the half drop windows receiving protection and a rectangular vision panel provided for each seat. In its later form, the vision panels became diamond-shaped, and centrally positioned on the plain side windows, although no such provision was made for the window on the nearside rear of the lower deck. Also of note is the transfer, sited below the used ticket box, proclaiming that 8 standing passengers could be carried at all times.

Far right The effect of bomb blast on an RT window. Passengers were advised to keep the winding window open during raids as this half was not fitted with netting. Small notices to this effect were produced for placing at the top of the winding windows. This one was left closed and only a few splinters of glass remain. The protected lower half has merely shattered, showing the netting's effectiveness in avoiding injuries to passengers. Forward facing windows had safety glass and were left clear.

Bus service needs had been affected by the war, and the need to economise led to some transfers of routes between garages. The October 1941 changes, which occurred on the 29th, saw Chelverton Road lose route 28 to Battersea (this route had not been officially worked by RTs since August 1940) and route 72 to Hammersmith (though it returned to Chelverton Road and RTs in March 1942).

Among the RTs transferred to Victoria was RT 1, and over the following months further vehicles moved into this garage, which included the last three to enter service, RTs 143 and 148 in December and RT 151 (appropriately the final one) in February 1942. Not all the vehicles entering service in late 1941 carried high stock numbers, for they included RTs 76-78, 82 and 91. Of these, RT 77 had actually been delivered to Chelverton Road on 12th July 1940 with RT 111. Because of the braking problems, both vehicles remained unlicensed and, two weeks later, were transferred to Potters Bar for storage, finally entering service on 1st October 1941.

By April 1942, with all the RTs in service, the buses were also being used on route 77A from Victoria garage. The Sunday allocation on route 137 also became RT, an arrangement which officially lasted for just six months.

To minimise the risk of injury to passengers from shattered windows, anti-splinter mesh was stuck on most of the windows of London buses during the latter part of 1940 and early 1941. Some RTs were damaged to a minor degree, though the class was fortunate at that stage in the war not to receive any serious casualties. Their enforced withdrawal from service with brake problems just before the London blitz may have been beneficial in this respect, but of the routes on which the RTs ran, only the 96 went near the worst-hit areas of the City and East End. At about the same time, another measure of wartime austerity was the reduction of the destination display to a single panel in what had been the 'via' box at front and rear.

As mentioned earlier, the 188 further RT buses ordered in November 1938, and for which the AEC list number XU1651 had been issued on 25th May 1939, did not materialise, being caught up in the disruption and then cessation of normal production that was occurring in many factories all over Britain as war needs took priority. Just what differences from the 2RT2 buses were intended for these vehicles remains a matter for speculation, as this and other XU lists are not thought to have survived. One obvious possibility was the revival of the 'tailless' chassis, as tried on RT 1, if it had been intended that these buses were to receive bodies of that type. This assumes that the Chiswick body workshops were to be adapted to construct metal-framed bodies, as clearly had been the original plan. Possible, but seeming less likely on balance, is the switch to the toroidal engine as fitted to RT 19 early in 1940 (described more fully in the next chapter) and to become standard post-war, as no such version of the 9.6-litre engine existed in May 1939. There are indications that AEC's engineers were swinging round to the view that this was the best line of development for the whole range of engines the company built, but they would have had to convince their opposite numbers at Chiswick, and there is no indication that the latter had been persuaded at that date to abandon the pot-cavity engine so strongly favoured since 1938.

In an effort to conserve fuel during the War, buses were permitted to lay over in London parks. Photographed in Hyde Park, RT 13 demonstrates that most famous of all 2RT2 features – the rear roof box – in use. One other 2RT2, an STL and some STs and LTs complete the line-up.
Hulton Picture Library

A famous shot of RT 25 at the Hendon 'Bell' terminus of route 28, extended to this point on Sundays until March 1942. The nearside mirror is just visible in its original position under the canopy and angled to allow the driver a view through the front lower saloon window. Also of note is the short lived, diagonally mounted mudflap below the platform. The photograph dates from the summer of 1940.
D W K Jones

Completion of the original order made it possible to produce an accurate figure for the construction costs of the 150 2RT2 vehicles. These, as reported for submission to the Board in October 1940, worked out at £2,011 per complete bus, of which £1,041 was for the chassis and £970 for the body. This was up on the original 2RT2 estimate of November 1938, and even more on Durrant's estimates of February 1938, due partly to the reduced number of vehicles compared to the original plan and partly to the interruptions of body construction due to delays in supply of materials and parts, pushing the cost per body up by £175.

However, in regard to the more modest rise of £91 each for the chassis, it is significant that reference is made to 'the general re-design of the chassis' as well as alterations in specification, adding automatic brake adjusters and chassis lubrication (evidently not in the original specification), and referring to the design of the rubber-mounting for the engine, the air brake system and improvements in engine design.

Even so, the chassis was still below list prices for much simpler double-deck models and the body less than the LPTB had paid for metal-framed bodies bought from Weymann, Park Royal and Leyland for STL and STD buses in 1936/7.

Among the closest comparisons of buses supplied to provincial operators were the AEC Regents supplied to Brighton Corporation early in 1939, which had A180 direct-injection 8.8-litre engines, preselective gearboxes of the standard D132 type and standard vacuum brakes. The chassis price reported by the municipality for these was £1,044, with £1,015 for Weymann metal-framed bodywork, these prices undoubtedly being influenced by keen competition to obtain that town's first municipal bus contracts on abandonment of its tramways; so on that basis, the RT figures seem good value.

The air compressor difficulties had been eased but not entirely overcome, for even the modified rotary units only ran for about 20,000 miles, equivalent to about 6 months' service, before requiring overhaul. This was just about tolerable to keep the buses in service but clearly unsatisfactory on a long term basis, especially on a vehicle where reduced need for maintenance was one of the main objectives.

The next step was the introduction of a heavier-duty reciprocating compressor which, fortunately, was a standard model being made by Clayton Dewandre for use on tanks and other military applications and thus in regular production. One of the last RTs to enter service, RT 143, was fitted with such a unit before doing so, though the chassis had a standard unit when completed by AEC in March 1940. The conversion is recorded as having been done on 29th November 1940, the same date given for the body being mounted, which rather suggests that this was a 'recording' date, not necessarily when either job had actually been done. The bus then spent a whole year at Chiswick, giving plenty of time for experimental work on it before entering service from Victoria garage on 1st November 1941.

It has been suggested that this and perhaps some other early installations at first retained the shaft-driven arrangement. If so, it seems likely to have been very short-lived indeed, even if the compressor was available in a suitable through-drive form, which is doubtful. The much more practical idea of driving the reciprocating compressor separately via a vee-belt drive from the front of the gearbox was adopted, in all probability from the beginning. This involved a minor modification to the underside of the body, and this led to the adoption of a modified code of 1/2RT for the chassis and RT2/1 for the body. Between March 1943 and November 1944 a further 49 buses were converted to this 1/2RT2/1 specification, most being done between December 1943 and April 1944. The unconverted buses soldiered on for the time being with the shaft-driven compressor and dynamo set-up.

Even the new belt-driven compressor arrangement was not entirely trouble-free, for the belt tension put quite a heavy loading on to the gearbox front bearing, its life being thereby shortened. A more radical change involving the balancing-out of this loading by taking the dynamo drive from the same source, but on the opposite side, was developed later, as described more fully in chapter six.

The two views of RT 133 were taken outside Victoria garage during the period from April 1942 and October 1943 while this vehicle was allocated there. In this case it had been in service from April 1941 after initial storage following completion in June 1940, with spells at both Chelverton Road and Putney Bridge, and had begun to look a little shabby, especially in regard to the white paint around the platform, very difficult to keep tidy. The front view shows the addition of a second headlamp by then fitted, and the small nearside mirror now sited to give the driver at least a limited view of passengers loading or alighting. The provision of a full-depth window at the rear of the platform instead of the shallower type usually fitted on rear entrance buses was another innovation pioneered by the type. The large white spot on the rear panel was originally applied to London Transport motor buses to help trolleybus drivers, unclear under blackout conditions as to whether it was permissible to overtake the vehicle in front if uncertain as to whether it might be another trolleybus. The Adamite non-slip edging to the platform was fitted to some members of the class when first built as a substitute for pyramid tread. At the end its life, this bus was to be the last 2RT2 bus in red livery in public service, when acting as stand-by for those painted green and operated from Hertford in 1955-56.
LT Museum

CHAPTER THREE

Finding More Customers

The only RT-type supplied to an operator other than London Transport in the period before 1946 was this example for Glasgow Corporation, intended to be displayed at the Commercial Motor Show planned for November 1939, but cancelled due to the outbreak of war. The bonnet and front mudguard design were to RT pattern, even to the inclusion of the pressing for the fleet number and the rather crude spring hook to hold the bonnet top down. Note that the front bulkhead window sill curved, the waistline of the Weymann body being slightly lower than on the London RT.
AEC/Gavin Booth collection

The philosophy followed by G. J. Rackham as AEC's chief engineer, and indeed during the whole of his career, was to standardise as widely as possible while yet offering operators the buses and lorries that suited them best. One of the key factors in this seemingly contradictory pair of aims was to use common parts and units in as broad a range of models as practicable – one standard spring shackle pin design fitted the entire AEC range over almost a quarter of a century, for example. However, another approach was to build chassis which could find as wide a market as possible without fundamental change, even though allowing some variations to suit individual preferences.

Thus the typical London bus was closely paralleled by 'provincial' models having much in common. This had been achieved very effectively in the 1930-31 period when the London ST and the Regent models sold to operators outside London differed only in relatively minor details. The standard STL of 1934-38 did not quite have the same universality, for Regent operators outside London were split in their preferences in regard to the type of engine and transmission. Even so, there was considerable 'commonality' among major components and indeed the whole collection of smaller items that went together to make up chassis, to spread the manufacturing costs very considerably, and thus reduce the costs with obvious benefit both to AEC and its customers.

The RT had struck out along a new path, and thus when introduced had fewer items in common with other AEC production models, as matters stood in 1939. At the same time, there were clear indications that it was regarded as a pointer to the way in which future urban bus design might be expected to go. London Transport, and the LGOC before it, was widely regarded as the world's most prestigious city bus operator, and thus any innovation it made was watched with great interest and had a good chance of setting a trend, at least among other British city fleets and in all probability in other parts of the world where British influence on such matters was strong.

The evidence rather suggests that the management at AEC had not quite got to the point of deciding to 'push' the RT, or a range of models derived from it, as the company's future bus standard by the time decisions were being reached as to what should be displayed at the 1939 Commercial Motor Show at Earls Court, London, planned for November of that year. Specific internal orders relating to it would have been in hand early in the year, when RT 1 had yet to appear and there was only the experimental running of the prototype chassis as ST 1140 as a basis for planning. That was clearly promising, but AEC's current range in 1939 was selling well and it seems more likely that the introduction of a new passenger range for general sale was in mind for rather later, possibly in late 1940, when it could have been introduced at the Scottish Show which alternated with Earls Court, or perhaps more probably for the next Commercial Show in London, which would have been due in November 1941.

AEC's range of types being produced was going through a period of some complexity, even if only the variations of Regent double-deck model are considered. The drawing office wallchart listed models which were regarded as standard in terms of production, and study of the 1939 version is revealing. To the outside world, AEC publicity gave the impression that the 7.7-litre engine was almost universal, but the 8.8-litre was shown on the chart as a production option in its A180 direct-injection form with either fluid transmission or crash gearbox. In fact, over 150 Regent buses so powered were built, mainly for municipal fleets such as Sheffield, Salford, Rochdale, Brighton and Doncaster, but also for others including Rhondda Transport Ltd and the Western Welsh Omnibus Co Ltd as well as for export to Sydney, in the 1938-40 period.

In addition, there had been the three STL buses (STL 2513-5) with engines of the experimental A805/1 type of completely redesigned 8.8-litre engine placed in service in April–May 1938 and the Regent demonstrator with similar engine, O6615413, registered JML409, running on hire to Bradford Corporation since 10th November 1938. Its chassis number suggests an earlier date, possibly late 1937, and hence possible initial use as a development vehicle – certainly it seems clear that it formed part of the heritage from which the RT sprang.

Another Regent having this 'half-way-house' engine, which had the redesigned crankshaft, bearings and timing gear as adopted for the A185 RT unit, in this case with a slightly different installation in some respects, was chassis number O6616956, delivered to Dundee Corporation in February 1940, numbered 112 and entering service registered YJ7586. The engine, A182-25, had the toroidal form of direct injection system, possibly the first toroidal A182. However, all of these A182 installations in Regent buses were based on the standard chassis and not the RT version, even though clearly reflecting similar lines of thought in the use of a compact larger-capacity engine which was less highly stressed than the 7.7 and could thus be expected to give longer service life before needing attention.

Thus the construction and bodying of a complete RT-type chassis for Glasgow Corporation, intended to be in time to appear at the 1939 Show, was very significant. In the event, the war caused the Show to be cancelled, but the bus was delivered in February 1940 (though reported not to have entered service until July of that year).

The Glasgow bus was given the chassis number O6616963, a few numbers on from the Dundee bus, and well after the London RT series O6616749-6899. However, AEC chassis numbers were issued on confirmation of order, and clearly the intention was to build it by the summer of 1939 to allow delivery to the bodybuilders in plenty of time for its completion before the Show. In view of the London RT chassis orders in hand, it was logical to build it along with early deliveries of them and indeed this appears to have been done, albeit later than intended.

AEC engine serial numbers generally give quite a clear indication of actual chassis build order and thus it is very significant that the Glasgow RT chassis had the first production A185 engine, number A185-1, suggesting even that it might have been the first RT-type chassis to come down the production line, that for RT 1 having been built in the experimental department. The chassis specification is understood to have been similar to that of the London 2RT production chassis and may have been virtually identical, apart from minor cosmetic features such as the use of an AEC rather than London Transport radiator badge and the chromium-plated 'soup plate' type of front hub cap fitted to many AEC passenger chassis in the 1938-9 period. It was not favoured by London Transport, save for what may have been a single instance on a trolleybus; Glasgow's was the only RT chassis to be so fitted.

However, a problem that was to recur in the early post-war period arose in the lack of provision of what AEC used to refer to as the 'structure', in other words the lower part of the cab, in those days usually supplied as part of the chassis on buses for general sale, but omitted on those for London Transport, which built the whole cab as part of the body. Other bodybuilders incorporated the AEC or other makers' structure into their body.

The Glasgow RT had the bonnet and nearside front mudguard fairing of standard RT pattern and it seems probable that these items were supplied with the chassis, perhaps made by London Transport to AEC order. The lower part of the cab may also have been based partly on LPTB drawings, a clue being given by the set of six louvres above the offside headlamp which matched the same number as those on RT 1 rather than the seven on a standard 2RT2 body. Other details of the cab were quite different and even though the lower edge of the windscreen was similarly curved, it did not have the rounded corners of the RT version, and the cab front was differently shaped to accommodate an offside headlamp. The head and side lamps themselves were of types widely used on non-London buses, with chromium-plated headlamp rims. The vehicle seems to have impressed Glasgow Corporation to sufficient effect that orders for 265 post-war AEC buses for this fleet, placed in service between 1948 and 1951, were for the basically similar Regent III model with 9.6-litre engine, preselective gearbox and air-pressure brakes, though not for the RT version.

This view of the Glasgow RT appears to have been taken sometime in the summer of 1940. The obligatory white paint had been applied to the front mudguards, which, incidentally are of the rubber type also fitted to most London 2RT2 buses when new. The cab design was unique to this vehicle, but the remainder of the body was as built by Weymann for Glasgow on standard AEC Regent and other chassis, with many typical Weymann characteristics, but special Glasgow features, notably the style of destination box. *Gavin Booth collection*

Just what might have happened had there been no war is a matter for speculation, but it seems virtually certain that this line of thought would have been taken further. Indeed, it is particularly striking that despite the war and all the demands for priority for military products, AEC engineers continued to spend considerable time on the RT, to an extent that went far beyond mere co-operation with Chiswick to overcome teething troubles with the 2RT buses. Admittedly the war was not yet at the stage when invasion of Britain seemed an imminent threat, and indeed the 1939-40 winter and spring were sometimes described as the 'phoney war', so little was the level of activity.

It seems very likely that in normal peacetime circumstances AEC would have built an RT-type chassis for its own use as a development vehicle, probably bodied for subsequent duties as a demonstrator, and very possibly Weymann might have been asked to build another body to something like the Glasgow pattern. Instead, RT 19 was hired back by AEC from London Transport, and in the event was to have quite an eventful career both as a basis for the further development of the type and as a demonstrator, spending more time on such activities for the rest of the war years than in normal passenger service. The charges for hire to AEC, incidentally, were based on a 10-year life for the bus, London Transport requesting an annual payment of one-tenth its original cost plus interest on the price it had paid.

RT 19 had run for only just over three weeks in service from Chelverton Road, following first licensing on 2nd January 1940, before being delicensed on the 27th of that month, on its return to Chiswick. The records indicate that it was next licensed on 19th February, its 'allocation' being shown as AEC, Southall, for use as a demonstration vehicle. The next entry, on 1st March 1940, includes reference to it receiving 'toroidal cylinder heads' and the rear axle ratio being altered to 5.2 to 1 from the 5.75 to 1 figure that was standard for the 2RT. It is also shown as being repainted 'light green' for use as a demonstrator and the unladen weight recorded was increased by 3cwt to 6tons 15cwt.

All this could not have occurred on that one day and other evidence shows that this was an entry recording what had occurred over the previous month. John Gillham, already an avid recorder of bus matters, was an apprentice at Chiswick works at the time, taking a close interest in what was going on, even though not free to go through closed doors into such places as the experimental shop. His records, which he has kindly consulted at my request, show that RT 19 appeared in the light green and cream livery of the Mansfield District Traction Company on 29th January 1940, the legal owner lettering at that stage being that of AEC, Southall, the vehicle then spending 'about a month' in the hands of the AEC experimental department at Southall. By 11th March he observed that the legal lettering had reverted to that of LPTB at 55 Broadway, despite the retention of Mansfield District livery.

The changes to the engine and axle ratio, the two inter-related, are particularly significant. An AEC experimental department report dated 2nd February 1940 makes it clear that the first 9.6-litre engine with the toroidal combustion chamber was actually a railcar unit, converted from a pot cavity engine, doubtless of type A182, produced for the GWR railcars then being built, though the report refers to a "second engine fitted to a London Transport vehicle", and it seems virtually certain that this was RT 19. At that date the bus is not recorded as having left Chiswick, where it had been since 27th January, but as it was delicensed it seems very probable that its move to AEC may not have been entered on the record card until it was relicensed for demonstration duties on 19th February.

The AEC delivery note for this chassis shows the engine number as having been changed from the original A185-27 to A185-15, the latter presumably 'borrowed' for the toroidal conversion and testbench running before being fitted to the vehicle. There is no date recorded for this change, but it is interesting to find that the change of axle ratio to 5.2 to 1 is dated on the delivery note as having taken place on 23rd February, four days after the bus was relicensed for its demonstration duties. This rather suggests that the change of ratio to take advantage of the modified engine's extra torque was something found to be desirable from road running rather than something that was initially planned. AEC's version of the toroidal combustion chamber gave more power and torque from a given sized engine with better fuel economy than the pot cavity used by Leyland.

No doubt there was a heightened consciousness of the need for fuel economy because of the threat to supplies of oil, in those days all imported, after the outbreak of war. The indications are, however, that AEC's engineers had made a long-term decision, as virtually all subsequent AEC engines, right up to the end of their manufacture, were to be of the toroidal type. The toroidal engine's increased output made it possible to use the higher gearing given by the change in axle ratio, itself helping fuel economy and, because the engine was thereby given a longer stride, helping to extend its life between overhauls. It is particularly interesting that this pioneer 9.6-litre conversion was not immediately sent back to London Transport for comparative trials with the other vehicles of the batch but sent off on a series of demonstration visits elsewhere. Clearly, by then AEC was convinced that this was the correct line of development, but there are grounds for thinking that this may not yet have been so at Chiswick.

In its basic mechanical specification, RT 19, as modified in February-March 1940, was effectively very close to the post-war RT (as well as to the standard 'provincial' Regent III as delivered to operators from 1947) in terms of performance and economy given by this combination of features, adding to the already radically new concept the RT offered in its overall design.

It duly went to Mansfield District for April and May 1940, beginning its career as a demonstrator which was to continue with a series of other operators for a period of just over two years. It returned to Chiswick from time to time during that period and John Gillham found clues as to its most recent movement from tickets on the floor or paper destination labels on the windows.

It had been common though by no means universal practice to paint buses in appropriate livery when they were visiting operators for a period of demonstration duty in public service, so the use of Mansfield green was in line with this. The idea of doing so in wartime is apt to seem curiously out of line with the image of general austerity associated with those years, but in that first winter many things were deceptively 'normal'. The green livery was applied in RT style, with the cream relief confined to the window surrounds and between-decks band, rather than following the layout with larger areas of cream normal for MDT, but the vehicle was lined out in the operator's usual style, a feature which looked rather out of place on the RT's modern outline. No attempt seems to have been made to alter the livery during the subsequent series of visits, but the war situation had worsened by then and such work might well have attracted criticism as being a wasteful diversion from the war effort.

After being taken out of service in January 1940, following only three weeks operation from Chelverton Road, RT 19 began what was to prove an extended career as a development and demonstration vehicle. Engine and rear axle ratio modifications brought it close to what was to be the post-war standard for both RT and 'provincial' Regent III models in these respects. It was repainted in the elaborately lined-out green and cream livery of Mansfield District Traction Company, modified in layout to follow the RT style, this operator being thought to have been the first to receive it on its demonstration tour, beginning in April 1940. It visited at least 22 operators in the period up to the summer of 1942, being seen here when operating for Nottingham Corporation on a wet day in September 1940. Note the use of the route number boxes to display the AEC triangle emblem – lettering in a side window described it as the "Regent" RT bus. *G.H.F. Atkins*

Operators who are known to have had RT 19 on loan during this period are listed below, with the approximate period of loan quoted where known. There may well have been others and it seems quite probable that, in addition, it may have 'called' at the premises of further operators nearby or en route to or from those listed, even if not operated in service:-

Mansfield District Traction Co	April to May 1940
Maidstone & District Motor Services Ltd	July 1940
Edinburgh Corporation	September 1940
Bradford Corporation	September 1940
Youngs Bus Service Ltd, Paisley	November 1940
Nottingham Corporation	Aug/Sept 1940 and Jan 1941
Sheffield Corporation	Mid-Feb to early March 1941
Leicester Corporation	March 1941
Enterprise & Silver Dawn Motor Services Ltd	May 1941
Wolverhampton Corporation	August 1941
Liverpool Corporation	October-November 1941
Manchester Corporation	November 1941
Rochdale Corporation	December 1941
West Bromwich Corporation	(date unknown) 1941
Glasgow Corporation	(date unknown) 1941
BMMO (Midland Red)	Dec 1941 to Jan 1942
Crosville Motor Services Ltd	January to February 1942
Birmingham Corporation	March 1942
Brighton Corporation	April 1942
Brighton Hove & District Omnibus Co Ltd	April 1942
Southdown Motor Services Ltd	May 1942
Luton Corporation	May to 15th June 1942

It was during this period that AEC began to refer to the new-generation version of the Regent model as the Regent Mark III. At first it seemed that this would be simply the RT with no more departure from the London Transport version than the Glasgow example, for early publicity was accompanied by illustration of London examples, or occasionally the Glasgow bus. Substantial and in most cases repeated orders for the post-war Regent III model were obtained. Most turned out to be in the model's 'provincial' form, with higher bonnet line and many minor changes, but they shared the then quite radical combination of 9.6-litre engine and air-operated preselective gearbox and brakes with the RT. This helped to spread costs as many parts were common or derived from the same castings or pressings. Major and/or repeated orders from undertakings known to have been visited by RT 19 were obtained from the following:

The Sheffield, Nottingham, Liverpool, Leicester, Rochdale, Brighton, Glasgow and Bradford Corporation undertakings, as well as the Midland General group of companies, which included Mansfield District. Smaller orders were obtained from Maidstone & District and the municipal undertakings of Edinburgh and Birmingham, the latter being among the limited group of operators which received RT-style chassis with the low bonnet in the immediate post-war period. Enterprise

& Silver Dawn was primarily a single-deck operator and standardised on the directly equivalent Regal III model in the post-war period. Some of these orders, as well as others for Regent III buses, are thought to have been placed during the period before 1942 when RT 19 was on its demonstration duties.

The vehicle was returned by AEC to LPTB Chiswick on 24th August 1942, and it received an overhaul that month, its classification at this point being reported to be 2/2RT2, although this is not shown on the vehicle's record card. Just what this code, if correct, signified is not clear, but it may have been a means of distinguishing the changes that had been carried out by AEC during the vehicle's absence, notably the March 1940 engine and axle ratio modifications, though it seems more than likely that the bus was also being used as an ongoing development vehicle by the AEC experimental department between its demonstration trips so there may have been other more minor alterations. Certainly AEC would have been keen to iron out any problems, such as those which had arisen with the brake system described in the last chapter, before allowing it to go out to operators as representing its future top-class model.

John Gillham's notes show the bus as having been repainted standard London Transport RT red livery in the last week of August 1942 though its records of movement show it as being put into storage at Victoria (GM) garage on 23rd September 1942 and the repainting red is not recorded until an entry of it being licensed, still at Victoria, on 1st October 1942. However, the card shows that the body was overhauled, starting on 24th August and completed on 23rd September 1942, and repainting back to standard livery seems bound to have been included in this process. On 3rd November 1942 it went to Putney Bridge garage to begin a period of normal service.

There can be little doubt that RT 19's status, incorporating the latest design specification as it then stood, was behind London Transport's choice as the basis for what was to be a further lengthy process of modification to become the prototype for the post-war production version of the RT, which was to be built in far greater numbers. AEC's experimental department had been sent the chassis of RT 52 on 22nd June 1944 when its body had been removed for repair after suffering air raid damage, and that chassis remained at Southall for nine months. Whether any work was done on it beyond possible repairs is not known, but it is significant that RT 19 was called in again. After running in normal service for a little over two years from Putney Bridge it was taken to Chiswick and is recorded as being delicensed on 19th March 1945. A week later the body, number 281, that it had carried from new and through its demonstrator career was removed, and the chassis sent to AEC once again.

By that stage, the war was drawing to a close, and victory was in sight. Indeed, a contract had been signed with AEC for the post-war supply of 1,000 RT chassis in April 1944, even before the main invasion of Europe to recapture it from the German occupation was launched in Normandy in June of that year. Clearly it was time to confirm precisely what their specification should be on the basis of operational experience.

This AEC advertisement was appearing in trade journals by the latter part of 1944. It is based on the picture of RT 53 shown on page 39, but with the lettering and fleet number touched out. It introduced the phrase Regent Mark III, not hitherto seen in print. At that stage there had been no public reference to a Regent Mark II, though some work had been done at AEC in 1936 on a double-deck O861 version, for Birmingham City Transport, of the lightweight Regal Mark II (model O862) introduced in 1935. In the event the O861 did not appear but its planned timing makes it feasible that the designation Regent Mark III was assigned to the RT project early in its development and that this is what the letters stand for. On the other hand, no reference to Regent III has been found in AEC documentation before the appearance of this advertisement, and when RT 19 was on its demonstration travels in 1940, lettering on the windows described it as the Regent RT bus. The use of Regent Mark II for the early post-war 7.7-litre model, which was almost identical to the standard crash-gearbox Regent of 1939, did not begin until late 1945. The impression was given that the post-war standard Regent III would be virtually identical to the RT. Among the features listed, most were to prove applicable to the standard 'provincial' model as well as the RT though Item 9, 'Greatly improved road visibility,' was only true of the RT, and the oil bath for the handbrake ratchet was dropped. The provincial Regent III had a bonnet level much the same as previous standard Regent models, something which caused great surprise when the first examples appeared in 1947.

Here is the Post-war Bus— THE "REGENT" MARK III

150 OF THESE VEHICLES HAVE PROVED THEIR VALUE IN SERVICE TO L.P.T.B. SINCE EARLY 1940

Salient Points of Improvement:

1. Passenger comfort
2. High-powered rubber-mounted engine
3. Finger-light steering
4. Pneumatic control finger-selected change speed
5. Air-pressure brakes
6. Fully automatic chassis lubrication
7. Exceptional drivers' comfort
8. Elimination of all manual efforts in control
9. Greatly improved road visibility
10. Oil-bath hand-brake ratchet.
11. Needle rollers on control shafts and pedals

Close-up view of driver's cab

CHAPTER FOUR

Winning the War and Planning for Peace

Below left Among the last vehicles to enter service, in March 1941, had been RT 147, one of those with an experimental gearbox. This photograph is of interest in showing it in the condition normal in the later war years, complete with restricted destination display, anti-blast netting on the windows and two headlamps. In this case, the latter are of the small-diameter type, originally introduced for military vehicles and standard on new 'utility' buses from about 1943 – for some reason the offside unit has no mask, which would mean it could not be used under the blackout regulations. The foremost upper-deck side window had been boarded over – many London buses ran for extended periods with some windows boarded. There was a glass scarcity, due to heavy demand as a consequence of air raids, but it seems that it might also have been concluded that this was not a high-priority task as long as there was a fair possibility of further damage.
W.J. Haynes

Far right Shortage of moquette led to some buses being re-upholstered in brown Rexine from 1944. RT 78 was overhauled in September of that year.
LT Museum 18560

The war years had taken their toll on London Transport's bus fleet, the normal high standards of body overhaul having suffered as Chiswick was required to aid the war effort. In addition to much of the fleet becoming shabby and ill-kempt, the timber-framed bodies used on all but a small minority of vehicles were often beginning to suffer from the effects of wood rot, not aided by the lack of proper repair and regular repainting. Even in 1945, well over 2,000 double-deckers were overdue for replacement by normal standards, with far fewer new buses than usual delivered during the war years, and most of them of the so-called 'utility' types on Guy and Daimler chassis built to wartime specification and having features which would not have been acceptable in normal times.

In addition, London had suffered over a longer period than elsewhere from air-raids, with many instances of extensive damage to buses. Most were repairable, but even abandoning the normal pre-war system of having a stock of spare bodies for most types of bus (to allow for the longer time needed for body overhauls as compared to chassis) did not provide an adequate number of spares to cope. No spare bodies had been built for the RT fleet, so when examples of this type became air-raid casualties, this method of overcoming the situation was not available at all. As it happened the RT class had not suffered any serious casualties during the heavy conventional bombing of 1940-42 but was not to be so lucky when the V1 flying bomb attacks began on 13th June 1944. RT 87 was withdrawn from

service four days later after being damaged in a V1 incident at Clapham Common, its body (357) removed for repair and being replaced by repaired body 299 from RT 52, the other casualty of about the same period, whose chassis was sent to AEC.

RT 66 had been the worst air raid casualty, its body having been so badly damaged by a flying bomb in June 1944 to be deemed beyond repair and scrapped. In due course body 357, ex-RT 87, was repaired and put on RT 66 in September 1944. In July 1944, RT 59 and RT 97, both badly damaged in similar incidents, were sent to Birmingham City Transport for repair, Chiswick being unable to cope with any more work of this kind at the time.

Among more minor repair jobs associated with war damage, RT 110 lost its front roof-mounted route number box, continuing to operate without it and giving a foretaste of the effect on the type's profile of the decision to abandon this item that came into effect on new production in 1948. Indeed it was used for comparative purposes together with RT 46 when the post-war blind layout was being reviewed. In 1954, towards the end of its service life, RT 110 was given a replacement roof box with a slightly smaller aperture than standard (see photo on page 102). Quite apart from bomb damage, normal wear on the moquette upholstery led to a decision to use a rather austere plain brown Rexine leathercloth to replace it on a few RT-type buses towards the end of the war. In addition, to avoid wasting paint, Chiswick adopted the practice of pouring small left-over amounts, principally of red, green and brown, into one large pot. In the later years of the war, the resulting varying shades of brown were used in place of grey on the roofs of overhauled buses, including the 2RT2s.

The damage suffered by RT 97 as a result of the nearby explosion of one of the unmanned V1 flying bombs was to set this bus on a new career, as described in Chapter Five. The V1 was one of Hitler's 'secret weapons', launched in large numbers against London in 1944. RT 97 was sent to Birmingham City Transport for repair but this had not been carried out when the bus was brought back to London, evidently because the sum allocated for its repair was judged inadequate. Apart from the damage to almost every panel, the distorted pillar angles indicate damage to the framing which would have required an almost complete rebuild. The impasse was overcome when funding for the 'Pay As You Board' experiment was added.

The original plan with regard to a prototype for the post-war fleet, decided upon in March 1945, was that one complete vehicle from the first 500 was to be purchased to act as a model for future production, the chassis from AEC but the body-builder not specified. In July 1945, however, this instruction was changed, cancelling the supply of a new chassis for this purpose but stating that that of RT 19 'which has been reconstructed by AEC to conform to the latest design' was to be used as the chassis for a new body to replace the one lost in June 1944 on RT 66, and authorising expenditure of £1,750 on the new body.

The selection of RT 19 for the prototype exercise was doubtless intended to speed up the arrival at a finalised and proven specification for post-war production. The work already done on this chassis, mainly in 1940, brought it close to the intended future standard in its basic features, though work incorporating further modifications and testing them was to keep it at Southall for over four months after its arrival there in March 1945. Apart from general updating of units, reverting to the short-tailed frame to suit the cantilevered rear platform as pioneered on RT 1 and to be standard for the post-war RT, the layout of the compressor and dynamo drive reached its final form. By driving both the compressor and dynamo from pulleys on the input shaft of the gearbox, one on each side, the resulting loads on the front gearbox bearing were roughly balanced. There is photographic evidence that this modification had been carried out on RT 19's chassis by July 1945.

However, a little of the time that the chassis spent in AEC hands in the summer of 1945 was used to allow members of the two leading bus operating industry trade associations, the Public Transport Association and the Municipal Passenger Transport Association, to inspect it in Red Lion Square when in London for their annual meetings in July. There was evidently still more work to be done, for following the return of the chassis to Chiswick on 3rd August, presumably for testing, it went back to AEC again, still in chassis form on 29th September. The war had ended in August with the surrender of Japan (the conflict in Europe having ended in May), and AEC was preparing to resume bus chassis production, though RT output was not due to restart until the following year.

Clearly, the prospect of obtaining a new body to post-war specification without further delay had proved as elusive as with the chassis. On 27th November 1945, RT 19 is recorded as being returned from AEC to Chiswick and receiving the body that it was to carry for the rest of its operational life. This was the original body from RT 1, number 18246, the reasoning behind this being that its basic design, with a largely metal-framed structure and the rear platform cantilevered from above rather than supported from the rear of the chassis, was nearer the proposed future body standard than the mainly timber-framed RT2 type.

There is some dispute as to just what the vehicle's coding was at this stage in its career. It has been claimed that it was recoded 2/2RT1 in November 1945, the change to the body coding being logical since the body from RT 1 had not been altered on transfer. However this does not appear on the record card, and it is a matter for speculation as to what coding the vehicle actually carried or should have carried at that stage.

What is remarkable is what the card did quote as the vehicle's code at this point, which was 3RT3. Admittedly the entry is not dated, but the style of handwriting is similar to others from around 1945, and it remained current until RT 19's original card was superseded by a new one, the space available to record transfers having been filled, with a final entry dated 3rd January 1948. Now 3RT3 was the code for the initial post-war standard version of the RT – the vehicles with fleet numbers RT 152 and upwards, of which the first examples did not go into service until 1947. No doubt 3RT3 would have been part of everyday language within the engineering department at Chiswick in 1945 as work on the specification proceeded, and since RT 19 was to act as the prototype for these vehicles, perhaps someone reasoned it was logical to apply it thus. Yet the post-war body was to differ from that on RT 1 fundamentally in its jig-built construction as well as in numerous details.

Moreover, it is a matter for speculation as to just how closely the chassis of RT 19 had been brought into line with the post-war specification at that date. The word prototype can be interpreted in two ways, one being that such a vehicle may be a working guide to the behaviour of the intended future production model, yet not necessarily conforming to the final design in ways regarded as less significant in assessing its performance. The other is that it should be precisely as was to be built in quantity, to confirm, so far as possible, that there would be no problems in any respect. The 1940 conversion of RT 19 to toroidal had clearly mapped the general direction for the future, but the post-war production vehicles had quite a number of changes in design, some of a kind which were bound to alter their behaviour from what had gone before.

A photograph taken at the time when RT 19's chassis was being viewed by members of the operator associations in July 1945 strongly suggests that the engine at that stage may have been as it had been since 1940, the rocker cover being of different shape to either earlier or later versions of the 9.6-litre engine. However, it seems possible that later in 1945 an engine more nearly approaching post-war A204 specification, and some changes to other units, such as the post-war types of fluid flywheel, J156 (18in, with baffle plate), and gearbox, D150 (with revised bearing arrangement), may have replaced the previous combination of modified A185, J150E and D140.

At that stage, body 18246 was still of 55-seat capacity, as had been necessary to meet the tight gross weight limit in force when RT 1 was new, but during a further period of almost two months at Chiswick it was modified to seat 56, the unladen weight being recorded when this work was done as 7tons 4cwt on 25th February 1946, creeping up towards the post-war standard of 7tons 10cwt.

The chassis of RT 1 was put on one side when its body was removed for transfer to RT 19 on 27th November 1945. It has been said that there were plans to re-use the chassis, and that the 2/2RT code was to have applied to it initially rather than RT 19. Indeed a photograph of what is clearly identifiable as RT 1's chassis has 2/2RT applied to it in drawing-office stencilled lettering. This suggests that the original plan might have been to use RT 1 as the prototype for the post-war RT by modifying its chassis to the latest specification, until it was decided that the chassis of RT 19, as it stood in 1945, was already much closer to what was wanted. After almost a year out of service, RT 19 was relicensed and went back to Putney Bridge garage on 1st March 1946.

Quite apart from the amount of development work already done on RT 19, RT 1 posed a problem because of its non-standard brake system, with some parts of German origin. Obtaining replacements was bound to be difficult and, even if this had been possible, such a course would have been politically unacceptable at the time had it become public knowledge. With one RT body short as a result of the severe air-raid damage to RT 66, and the potential of good use of the standard chassis items from RT 1 for spares, it doubtless seemed logical simply to dismantle its chassis, sad though such an early end to a truly historic vehicle was. It does seem that the matter remained in doubt for a time, for it was not until 4th September 1946 that it was recorded that this had been done.

With the return of peace and new RT deliveries expected to begin arriving within a year or so, the 2RT2 buses moved into a new era, when they would no longer be the Board's best and brightest stars in its fleet. Their story was far from over though, even in terms of new developments.

After seven months in service from Putney Bridge following the major work done in 1945-46, RT 19 was withdrawn on 1st October 1946 to spend three weeks at AEC, this time with body still in place. Just what this visit was for is not clear – possibly no more than detailed assessment of how everything was behaving in terms of wear rates, etc, though clearly unit changes to bring it closer to post-war standard would have been possible if a little unlikely after so short a period in service, particularly in view of what was to happen in 1948. By that date, deliveries of new 3RT chassis were well under way, though none was in service until the following year.

The bus went back into service on 24th October 1946, this time from Chelverton Road, and ran until 1st January 1948, when it was withdrawn again and, with body removed on the 6th, returned yet again to AEC. It was at this point that a second record card was opened, and the initial code on this duly appears as 3RT3, doubtless simply copied from the previous one. However, it then changes (undated and in different writing) to 2/2RT1, which is what it could be said to have been from 1945, and then again, in what appears to be the same writing as the 2/2RT1 entry, to 3RT1 – almost suggesting that someone was catching up with reality.

Another significant point on this second card is that the axle ratio entry begins at 5.2, the figure that had applied since 1940, but is then crossed out (not usual practice on these cards) and 5.166 appears. This latter ratio was that of the post-war 3RT, so slightly different to have negligible effect on performance, but reflecting an internal redesign. It thus suggests that the axle had been changed, and thus it may be that RT 19's chassis had finally become so close to the 3RT specification as to truly justify it having the same code. Just possibly RT 19 was fitted with a full set of production 3RT units at this point, and clearly this would be an easier task if the body was removed.

Details of the fuel tank mountings on the nearside and air system items on the offside show this to be the chassis of RT 1. The condition, clean but certainly not freshly painted, suggests that it was taken when the body was removed after having been in service – it is noteworthy that the compressor and dynamo remain shaft-driven, as built. Another interesting detail is the use of a 'mushroom' pedal, similar to the accelerator, for the gear-change pedal; the chassis had a conventional pedal when photographed at AEC in March 1939, and the version shown, reminiscent of similar practice on some early AEC Q-type chassis with vacuum gearchange, was not adopted as standard. The edge of the photograph was inscribed 2/2RT in drawing office stencils, implying that it was a Chiswick record picture, and lending support to the view that at one stage it had been proposed to update RT 1, instead of using RT 19, as the prototype for the post-war version of the RT.

CHAPTER FIVE

RT 97 and RTC 1

Towards the end of the 1939-45 war, there was a readiness to consider fresh ideas alongside the wish to soon be able to restore standards of public transport to peacetime levels. It may have been partly a by-product of observations reported by transport men in the forces in such European countries as Italy and then France as well as Germany, as the Allied invasion progressed, that caused an interest in unfamiliar methods of fare collection.

The idea of passengers paying immediately on entering the vehicle excited considerable interest. The reason behind it at that stage was not the elimination of the conductor, as eventually made such methods usual in Britain in more recent times. The principle as then favoured was based on the idea of the conductor collecting fares while seated at a desk near the entrance, from which passengers would then pass to their seats.

Just why such an idea seemed attractive by comparison to the traditional mobile conductor collecting fares from passengers at their seats is not, at first thought, very easy to understand, half a century later. It was probably to do with the overcrowding which was suffered on most buses in wartime, and was to remain a problem well into the post-war years. This caused great difficulty for conductors in getting round the bus to take fares, resulting in a greater percentage of passengers not paying, which naturally concerned operators and was thus probably the major influence. Related to this was the idea that the congestion and consequent delay on a crowded bus of normal British design, when passengers wishing to board had to wait for those struggling to get along the crowded gangway to get off, might be reduced if the entrance arrangements were revised.

London Transport decided to investigate the possibilities by modifying existing vehicles and using variations of design to explore the pros and cons of each. There were five in all, the first being STL 1793, which was converted to centre-entrance/exit layout in August 1944. The seating capacity remained at 56, though rearranged internally, but delays at stops when more passengers wanted to board than could be accommodated in the standing space near the ticket-issuing desk soon proved to be a problem. Trolleybus 61, dating from 1933, was the only vehicle in the fleet already having a centre entrance and was converted to 'Pay As You Board' (PAYB) layout in March 1945 – this was laid out rather more spaciously. A different approach was tried on STL 2284, with a front entrance and a centre exit, both quite narrow (the seating capacity being reduced to 50, though with four additional tip-up seats on the platform), which took to the road in this form in November 1945. Another trolleybus, basically standard 378, was fitted with platform doors to its rear entrance and rearranged staircase in December.

The RT class came into the picture when RT 97, which had been out of use since being damaged in an air raid in July 1944, was selected as the last vehicle to take part in the PAYB experiments. It is believed to have been this prospect that caused it to be brought back in December 1944 from Birmingham City Transport's central workshops, where it had been sent for repair, with the work not done. However, with the RT as the Board's chosen bus type for the future, its conversion may have been further delayed to gain the benefit of operating experience with the conversions of STL buses and trolleybuses before deciding on its form.

It was decided to retain the rear-entrance layout, so outwardly it differed much less from standard than the STL conversions, though platform doors were fitted. This in turn meant that provision for emergency exit from the lower deck had to be made, and in fact two such exits were fitted, one being a door directly opposite the entrance while the rear platform window was hinged to open in a similar way to the upper-deck emergency window. However, the overall extent of the work done was much more than the mildly altered external appearance suggested.

Initially, RT 97's PAYB conversion was designed to allow it to run as a Central Area bus from Kingston garage on route 65 from January 1946; two earlier PAYB conversions of STL-type buses were already at work on this route. Unlike the latter, the rear-entrance layout was retained, but quite extensive work was done to fit air-operated sliding doors, add the conductor's desk and move the staircase to gain more room on the platform. The livery at that stage was as had been used for 2RT2 buses since their operation began.

Interior views of RT 97 after conversion to PAYB layout, showing the modifications needed at the rear of both saloons in connection with the installation of a ticket desk and enlargement of the platform area by moving the staircase. The level of the platform was raised to that of the lower saloon floor.

The entrance doors were of the sliding type with two sections, of which the rearmost slid forward and then picked up the forward section as it moved, both then sliding alongside each other to occupy a space just behind the rear mudguard. The snag with this arrangement was that the resulting doorway opening was significantly narrower than the standard platform opening. The doors were powered by the standard RT air system and suspended from the body structure above, which was reinforced accordingly, as was the platform. The seating capacity was reduced to 50, there being seats for only 20 in the lower deck as the longitudinal seats over the rear wheel arches were replaced by the conductor's position and stairs – upstairs the total number was unchanged but differently arranged, with a full-width bench seat at the rear behind the repositioned staircase. In this form, the unladen weight was increased to 7tons 1cwt.

There is some uncertainty as to what code the vehicle bore at that point, one version suggesting it was 1/2RT2/2 – the chassis standard, with the 'first stage' belt-driven compressor modification applied to this vehicle in April 1944, three months before it received the bomb damage that started its unique career, and RT2/2 to signify the PAYB body changes. The problem with this is that RT2/2 was used to signify the changes to bodywork associated with the second-stage compressor and dynamo drive change which caused the chassis code to become 3/2RT and applied to 99 buses, though not until 1949-51. Could RT 97's brief existence in this form have been forgotten by then? In theory, LT's systematic approach should have made this impossible but other lapses did occur, as we have seen. On the other hand, theories that RT 97 was coded 9RT10, or even 5RT5, when converted in 1945 to PAYB form seem completely illogical having regard to what was running, on order or even planned at that date. The chassis was still standard and the body, though modified, still clearly basically an RT2.

The work was completed in December 1945 and on the 2nd of the following month it went to Kingston garage for operation on the 65 route from the 4th, joining the two PAYB STL buses that were already running, though other buses on the route were much older petrol-engined ST and early STL types with crash gearboxes. Despite this, RT 97 consistently ran late just as STL 1793 and 2284 were doing. The basic problem with all three was that, at busy stops, the bus could not move off until all those waiting to board had been able to join those already queuing to pay their fares to the conductor on the platform of the bus, and there simply wasn't sufficient space for enough of them to be accommodated.

In addition, the passenger reaction was negative, for people could see no justification for being made to stand, often still outside the bus exposed to the weather or, if on board, hanging on as it moved off, with money ready, waiting to pay their fare when other buses had conductors who did at least allow passengers to sit down before paying their fares.

The conductor sat at a desk over the nearside rear wheel arch with, in front of him, a bulky ticket-issuing machine made by National Cash Register, and very like the old-style mechanical cash register machines still widely used in shops at that date. On the offside, a completely new staircase was arranged to run forward over the offside rear wheel arch, leaving almost the whole of the platform clear for passengers. This was primarily to give space for them to stand while awaiting payment of their fares, but it gave a spacious entrance layout reminiscent of that on the 1930-31 ST and LT type buses.

This nearside view of PAYB RT 97 shows the reduced depth destination display made necessary by the addition below it of platform door gear. Interestingly the route number is omitted, despite there being no provision for a nearside route number plate. The emergency valve for the platform door can be seen to the left of the rear wheel. *LT Museum*

Modifications to the offside included the addition of an emergency exit and a reduced width window next to the repositioned stairs. *LT Museum*

RT 97 and RTC 1

It was decided to try PAYB on Green Line duty, and RT 97 officially became RT 97c, barely altered in design, but smartly painted in two-tone green livery then newly introduced for vehicles on such duty. It reappeared in this form on route 721 in April 1946, and is seen here at Romford. Although the PAYB fare collection was again unpopular, the vehicle was well-liked for its comfort by comparison with the utility Daimler double-deckers then operating the route. *A.T. Smith*

This period of experimental running lasted less than three months, the vehicle being taken out of service and sent back to Chiswick on 25th March 1946. However, the bus was not out of service for long, for the PAYB experiments moved to the country area, the reasoning being that the delay problem was associated with busy routes. RT 97 was selected for use on a Green Line service, and was repainted in the then newly established post-war two-tone green livery adopted for such vehicles, but with bronze lining-out – it also had the benefit of the front wheel discs originally standard on the 10T10 type of Green Line coach (which also were fitted to RT 4 and RT 39 for the victory parade a couple of months later). The overall effect of this, the first RT to appear in a London Transport green livery, was very smart.

The only practical modification was the addition of a small sliding panel in the window over the nearside wheelarch to provide ventilation for the conductor. The route selected for the experiment was the 721 running from Aldgate to Brentwood and operated from Romford (RE) garage, the vehicle arriving there on 18th April 1946. At this time it was officially coded RT 97c, the suffix being a hangover from the pre-war practice of giving Green Line vehicles this form of identification. The suffix letter appears never to have been applied to the vehicle however. The 721 was a busy route, which had been reinstated the previous month, using Daimler CWA6 double-deckers. These were hardly coach-like, their bodywork being built by Duple to the wartime Ministry of Supply 'utility' standard, admittedly in its final 'relaxed' form with upholstered seats and slightly less severe outline. Their main saving grace was that they were new, their standard of comfort being well below normal Green Line standards and indeed inferior to a standard late pre-war STL.

By comparison, RT 97c was luxurious, and indeed the standard London Transport bus seat was not greatly different from the Green Line version, both being shaped to give good support. The vehicle was popular on this type of duty but, again, the fare collection system was not, the main factor being again the problem of standing on a moving vehicle aggravated in this case by the fact that more passengers had luggage. There was little point to the PAYB system using a seated conductor on a Green Line route, when there was plenty of time for a conductor to move round the vehicle to collect the fares in the normal way. After ten weeks, the PAYB experiment was abandoned, but the vehicle remained on the route until 5th January 1947, almost exactly a year since it had first appeared in PAYB form, when it was again withdrawn. It was taken back to Chiswick the following day.

Left A nearside view of RT 97 in Green Line form at Aldgate. A sliding window has been provided behind the conductor's seat. *LT Museum*

Facing page Another scene at Aldgate, demonstrating the slower boarding of the PAYB buses. *LT Museum*

Above and facing page Much attention was given by the designers of RTC 1 to its bonnet and radiator grille design, as shown in these photographs of their model, which underwent a number of modifications before a full size mock-up of the vehicle was built at Chiswick. Note the use of 5RT5 as a 'fleet number', clearly indicating that this code had been chosen from the early stages of the project.

This time, an altogether more ambitious scheme was in mind, though it may well have been the success of the vehicle on Green Line duty that led to its selection as the basis for what amounted to a new concept of Green Line double-decker. The idea had fascinated the engineers at Chiswick since almost the beginning of the Green Line network of limited-stop services in 1930, for it soon became clear that some of them attracted enough passengers to justify a larger vehicle than the single-deckers with which all routes were run at first. Yet comfort was also a part of the attraction, so under normal circumstances something better than a standard double-decker bus was wanted.

The first attempt had been LT 1137, an AEC Renown six-wheeler, on which quite an exotic double-deck coach body was built at Chiswick in 1931. That remained unique, as did Q 188, the only six-wheeled example of the AEC Q side-engined model, which had bodywork to LPTB specification built by

Park Royal in 1937. The idea of the double-deck coach was not pursued further, though use of standard STL double-deckers on Green Line routes was a feature of wartime operations for a relatively brief period.

Clearly the post-war situation, with more demand for travel than ever, reactivated the line of thought; and the fact that double-deckers were being operated on some routes despite their unsatisfactory design underlined the point.

Official authority for the project, ultimately to emerge as RTC 1, was given in October 1947, at the same time as that for a 'chassisless' (i.e. integral-construction) double-decker using aluminium for the structural framework – the latter being the first stage in the development of the Routemaster, though that was a longer-term project. Even so, the association of the two projects in their early stages was significant, for some aspects of the two were alike, and RTC 1 acted as a test-bed for certain ideas incorporated in RM 1.

In other times, a brand new prototype would probably have been built, but with immense pressure on getting production of standard RT buses under way (only about 85 of the 1,000 of the post-war type ordered in 1944 having entered service at the time the RTC project was sanctioned) a rebuild was judged to be the best way forward. Indeed, it seems clear that this was what was in mind at Chiswick when RT 97 was withdrawn at the beginning of 1947. The whole exercise was carried out by the experimental department and, ironically, one senses that it was deliberately made to look as unlike a standard RT as possible, perhaps as a sort of mental compensation for the fact that it was a rebuild of one. It was a slow process, most of the first year from authorisation being occupied by design study. Both body and chassis were stripped down to the basic frame structures and the official record notes 'overhaul' to have begun in June 1948. The vehicle was briefly demonstrated to the technical press in January 1949.

Right The bonnet design on the full size mock-up was the one used on the finished bus. By this stage, the angled bottom edge of the driver's windscreen has also entered the design. The standard width waistband and provision for the route board to be fixed below it, rather than on it, were the only parts of the mock-up not carried forward unchanged to the RTC as built.

Most obviously, the frontal appearance was quite different. Gone were the familiar AEC radiator and bonnet and in their place a rounded bonnet outline, lower at its highest point than even the RT standard (to a mild degree not unlike London Transport's 1939 TF underfloor-engined coach, which had a half cab and rounded 'bonnet'), though in this case with a wider low-level grille. The profile was also altered, with less of the well-rounded effect of the RT, the cab front sloping all the way down to the point where it merged into the the grille panel, which also incorporated the headlamps and a foglamp. Car styling was altering quite markedly at that time, and there was a distinct resemblance in the grille design to contem-

porary trends exemplified by, for example, the Morris Minor as introduced in its initial post-war form in 1948. The trade press reported that the shade of green used for the exterior of RTC 1 was darker than that normally seen on Green Line coaches. A nice touch was a fluorescently-lit Green Line name on the light green waistband front and back.

The radically altered frontal appearance was made possible because of the removal of the radiator to a position under the stairs which, incidentally, were moved back to the conventional RT position in the course of the rebuild. The radiator operated as a heat exchanger, with powerful electric fan, which formed part of an elaborate heating and ventilating system, claimed to

allow the use of the whole waste heat output from the engine if required for interior heating, the heated air being carried by ducting at floor level into the two saloons. In summer, the fan discharged the air heated by passing through the heat exchanger to the rear of the vehicle.

The strong effort made to make RTC 1 look quite different, both externally and internally, from the standard RT was despite the widespread admiration the latter was attracting. It is sometimes said that the sincerest form of flattery is imitation, and echoes of the RT were to influence double-deck bus design from a variety of makers well into the 1950s. Instead of relying on London Transport's well-proven skills in functional vehicle design, outside consultants Norbert Dutton and Douglas Scott, belonging to the category just beginning to become known as industrial designers, were engaged. Among the consequences, RTC 1 had slim intermediate chromium-plated window pillars, lacking the radiused corners of the standard design. The opening windows were of the full-drop type, winding down into the body sides but with the movement limited to give a similar opening to the standard half-drop. The effect had clear resemblances to contemporary practice on production coaches as built by such specialist concerns as Duple, Burlingham or Harrington. None of the front windows, upstairs or downstairs, opened. All windows opened on each side except the first and last on both decks.

The RT basis of the vehicle remained clearly evident from the rear, where the characteristic rear dome and emergency exit continued to be easily recognisable, quite apart from the FXT272 registration – incidentally the body number 366 also remained unchanged, being as originally mounted on RT 97 in April 1940. The rear entrance had a two-step layout and in some respects, such as the door operating gear, the vehicle inherited part of its design from the 1946 PAYB conversion.
LT Museum U45987

Facing page left Removal of the radiator improved access to the front of the engine, but the built-up nearside mudguard made items further back more difficult to reach. To give adequate performance, the engine was a toroidal-cavity unit set to develop 120bhp, but the increased unladen weight of 8 tons more than compensated for this. *LT Museum*

Facing page right Interior, lower deck. The floor was covered in a mottled mid to light green rubber material, which must have been rather more difficult to keep clean than the conventional flooring in RT 97. The special seat frame design incorporated foot rests for the passengers sitting in the next row back. *LT Museum 102/7*

Within, the seats did not have the neat style of metal frame as used on London Transport's standard types of buses or coaches since 1936, the framing being concealed within rather square-cut though comfortable-looking moquette-covered seats. An unusual-looking cranked type of grab rail was fitted at the top of each. Fluorescent lighting was quite novel within public transport vehicles at the time, the units being formed into continuous strips running along the cove panels in each deck. The overall colour scheme within both decks was green in varying shades below the waistline and cream above, a little gloomy in general effect.

The seating capacity was now 46, with 26 on the upper-deck and 20 below. A legacy from the PAYB experiment was the provision of a seat for the conductor, but only for use when not circulating around the vehicle collecting fares in the normal manner. The modifications from standard led to the vehicle being recoded 5RT5, 4RT4 having been used for an 8ft-wide version of the RT which never got past a mock-up stage. As rebuilt, the unladen weight went up to 8 tons, which was 1ton 8cwt above the original figure for a standard 2RT2 bus, and heavy for a 26ft by 7ft 6in double-decker of any type.

The engine remained of basically original type but was converted to the toroidal cavity combustion chamber pattern, being classified as A185B and the power output quoted as 120bhp, a little over the London Transport standard of 115bhp for a toroidal 9.6-litre unit, doubtless to compensate for the extra weight. AEC records show that A185B was listed as a London Transport 'spares' engine type, supplied less flywheel and fuel injection pump, which may imply that it was an engine made up at Chiswick from spares stock. No alteration in rear axle ratio from the standard 2RT figure of 5.75 to 1 was made, so the governed maximum speed would have been slower than a standard post-war RT.

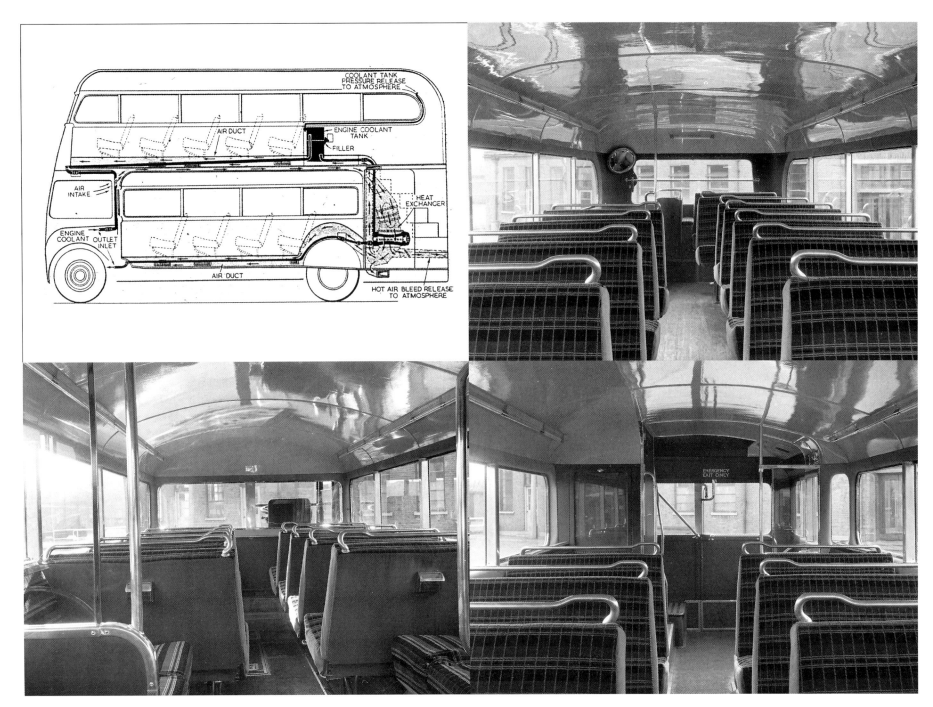

Top left The way in which air, warmed by the rear-mounted heat exchanger in winter, was circulated round the interior of RTC 1 is conveyed in this diagram. Also marked is the water circulation system, taking engine coolant to the header tank on the upper deck and then down through the heat exchanger and back to the engine. With the wisdom of hindsight, it is not difficult to see how trouble could arise from air locks building up in the water system – similar problems were not uncommon on other later designs of double-decker also having water circulating heaters or heat-exchangers at upper-deck level, until the need for care in eliminating sources of leaks, often of air into the system, was realised.

Other photos Interior views of RTC 1 before entry into service. *LT Museum*

LONDON TRANSPORT

GREEN LINE EXPERIMENTAL
DOUBLE-DECK COACH

LONDON TRANSPORT invites comments from passengers upon this experimental double-deck coach.

Many features designed to make your journey more comfortable have been incorporated so that they can be thoroughly tested in service, and your attention is specially directed to the following :—

★ Luxury type seating

★ Supple vehicle springing

★ Fluorescent lighting

★ Extra-wide windows

★ Improved visibility

★ Luggage space beneath seats

★ Novel system of heating and ventilation which, incidentally, renders smoking possible on both decks

PLEASE WRITE TO THE PUBLIC RELATIONS OFFICER, LONDON TRANSPORT, 55 BROADWAY, LONDON, S.W.1

Far left A leaflet issued by London Transport seeking views from passengers about RTC 1.

Left A tip-up seat was provided for the conductor to use when he was not busy. It was in front of the sliding door pocket by the entrance, photographed here through the open emergency exit.

Below Among the Green Line routes on which RTC 1 operated was the 711, on which it ran for just over two weeks in May 1949. *Michael Rooum*

After its two years off the road, the vehicle emerged from Chiswick for a press demonstration in its new guise as RTC 1 in January 1949, though the process of testing and adjustment was still incomplete and it was not until 31st March that it was licensed for the beginning of a series of trials on Green Line routes. Initially it was allocated to Windsor, but moved to Hertford before entering public service on 6th April on the 715 route. It moved to Reigate on 2nd May to work on the 711 and then on the 18th of the same month to Hemel Hempstead for the 708 before returning to Windsor, this time to go into service on the 704. Then it went back to Hertford on 25th July for a longer spell on the 715, continuing until November.

Not to put too fine a point on it, the end result of the exercise was a failure. The elaborate heating and ventilating system, with its repositioning of the radiator away from the natural air flow at the front of the vehicle, led to repeated overheating of the engine. This result hardly comes as a surprise when one remembers the trouble, often centred on overheating, that was to be experienced with many of the early rear-engined buses in the 1960s, and should perhaps have been taken as a warning of the problems likely to be met with rear-mounted radiator layouts generally, unless very carefully designed. Unreliability is a failing that is impossible to tolerate on a bus or coach, but the vehicle was unpopular in other ways.

Although new high-deflection road springs designed to suit the vehicle's weight distribution had been fitted and anti-roll bars added on both axles, the ride was criticised, with reports of a swaying motion and pitching, apparently worst when accelerating. The standard post-war RT did not have dampers, a surprising omission on what was in most ways a 'state of the art' design, perhaps reflecting the generally good state of London's streets in those days and maybe the limited opportunities for relatively high speed on a London bus route. The RT could develop a fairly serious pitching motion if driven fast on an uneven road, and it seems possible that the modifications to RTC 1, in particular the transfer of more weight from the front to the rear, may have worsened matters despite the suspension modifications. There was also criticism of the lack of the overhead racks for coats and luggage found on the single-deck Green Line fleet at the time. So the idea of a true double-deck coach for such duty was shelved once again, though it was decided to replace the Green Line D-class Daimler CWA6 double-deckers based at Romford with new RT vehicles, to standard specification apart from their livery, in 1950.

Right Demoted to country bus work at Leatherhead, RTC 1 spent most of its time on route 468 at first but was further humbled by being moved to the lighter 416 route. Here it is seen in its final condition with full country bus livery. *Alan B. Cross*

Facing page upper Upper deck interior after its 1952 overhaul, when its fluorescent lighting was replaced by tungsten. *John Gillham*

Facing page lower Photographed in 1955 just six years old and after London Transport had no further use for it, RTC 1 looks very sorry for itself at the premises of Lancashire Motor Traders, Salford, and ready for scrap. In fact, its potential was seen and work was done on it to make it suitable for further use. RTC 1 spent five years between 1956 and 1961 being employed as staff transport for the football pools firm Vernons of Liverpool. Ironically, this was its longest spell on a single type of duty at any stage in its career. *John Fozard*

Meanwhile, RTC 1 was moved to Leatherhead on 28th December 1949, after just under nine months in Green Line service, and demoted to use as a country bus. It lost its platform doors, Green Line lettering and the light green relief colour and now had a country bus cream waistband and upper deck window surrounds. It ran on local routes 468 and 416 until March 1952, when taken out of service for a light overhaul, during which its fluorescent lighting was removed, conventional seat cushions were fitted and the livery became the standard country green with just the between-decks band in cream. It then ran, again from Leatherhead, for a little less than a year in this form before being withdrawn from service on 1st March 1953. It was still proving unreliable and was costing too much to keep in use.

It returned to Chiswick, where it had a final spell as an experimental department vehicle, being used for further work on unorthodox engine cooling and heating systems. It must be remembered that by this date the other project authorised at the same time back in 1947, the integral double-deck bus, was now at quite an advanced stage and would appear as the prototype Routemaster the following year. It too had a radiator removed from its normal position as built, and no doubt RTC 1 helped to provide useful data for its design. In a sense, it may also have influenced it in a way that might seem negative, for the reports of poor riding may well have encouraged the decision to adopt independent front suspension for the Routemaster, giving ride quality that was greatly improved. There was also the involvement of Douglas Scott in its internal and frontal external appearance, early schemes for the latter having a distinct 'echo' of RTC 1, so it could be said to have been more influential than might at first appear, even though unsuccessful in itself. Ultimately, the idea of production double-deck Green Line coaches did materialise, in the form of the RMC and RCL Routemaster variants of 1962 and 1965, so the concept did prove basically sound after so many years of unsuccessful attempts.

Ultimately, with RM 1 complete and running, when W. North & Sons of Leeds, the well-known dealers in old buses, made an enquiry as to whether RTC 1 might be for sale, a deal was eventually done, and the vehicle left Chiswick for the last time in March 1955. Even so, its future looked dubious as various essential items including the radiator and dynamo had been removed because they were still useful from a development viewpoint. After sale to another dealer, Lancashire Motor Traders of Salford in June of that year, some work was carried out on it and the vehicle eventually found a home with Vernon's, the football pool concern, based in Kirkby, Liverpool, in July 1956, remaining with them for staff transport until sold to another concern, Pitbow Ltd, of Sandwich, Kent, in August 1961. It does not appear to have been sent to its new owner, however, and shortly afterwards met its end on Merseyside when scrapped by W. Price, dealers, of Birkenhead.

CHAPTER SIX

2RT2s in the Post-War Years

The aftermath of the war was still evident for some time after hostilities ceased and indeed some changes it brought about were permanent. So far as the 2RT2 buses were concerned, there were no spare bodies and serious damage resulted in vehicles being off the road for quite long periods. Of the two buses sent to Birmingham City Transport's sizeable and well-equipped Tyburn Road Works for repair after bomb damage in

August 1944, RT 59 had returned repaired in December and was relicensed for service from Chelverton Road in January 1945. The other, RT 97, was also taken back in December 1944 but had not been repaired and was to spend a further year out of service, though this was partly because of the first stage in a sequence of events that have been covered in the previous chapter.

A few attempts have been made to bring RT 24 to peacetime condition with some of the anti-splinter mesh and the offside headlamp mask having been removed, as can be seen in this view of the vehicle taken outside a boarded-up bank in Esher High Street during 1945. RT 24 was one of the small number of 2RT2s that remained in service throughout the war years, apart from a short overhaul lasting just two weeks during the spring of 1943.

At the end of December 1945, the routes to which 2RT2s were allocated could be summarised as follows (types used by other garages are shown in brackets):

14   Putney (Oxford Road) to Hornsey Rise (Daily). Monday to Saturday journeys extended to Kingston. Operated by Putney Bridge and Holloway (STL).

30   Hackney Wick to Roehampton (Earl Spencer) (Daily). Operated by Chelverton Road and Hackney (STL).

37   Peckham (Lord Hill) to Hounslow garage (Daily). Operated by Chelverton Road, Nunhead (LT) and Mortlake (LT).

72   Esher to East Acton (Ducane Road) (Daily). Extended on Sundays to North Wembley (Preston Hotel). Operated by Chelverton Road and Alperton (Sundays, G).

74   Camden Town to Putney Bridge (Green Man) (Daily). Extended Monday to Saturday peak hours and Saturday afternoons to Putney Heath (Telegraph Inn). Operated solely by Putney Bridge.

85   Putney Bridge station to Kingston (Daily). Operated solely by Putney Bridge.

93   Putney Bridge station to Epsom station. Additional service on Mondays to Saturdays and Sundays works journeys from Putney Bridge to North Cheam (Priory Road) (Daily). Operated by Putney Bridge and Sutton (STL).

96   Putney Common (Spencer Arms) and Wanstead (Red House) (Monday to Saturday). Operated by Putney Bridge and Forest Gate (STL).

Clues to the events of the war years remained for long afterwards. RT 59, badly damaged by enemy action in 1944, was repaired by Birmingham City Transport's Tyburn Road Works when Chiswick was unable to deal with it. That establishment had very high standards, but a small difference in appearance of this vehicle gave a clue to different hands being at work, as the valance under the nearside canopy, normally deeper on the RT2 body than on most other RT variants, was cut back, and the radius at the corner nearest the windscreen was more prominent than usual. The front mudguards appear still to be of the rubber type, prone to lose their shape in later years.

The unavoidable neglect of the war years took its toll on the appearance of the 2RT2 buses, this view of RT 16, also in service on the 30, being fairly typical. There are several creased panels, the cover giving access to the automatic lubricator no longer fits properly and the rear wheel disc is missing – this last feature often tended to persist, even after overhaul. The photograph probably dates from the summer of 1946, certainly before September of that year, when an overhauled body was fitted – RT 16 was one of the minority of buses of this type to receive a different body in the course of overhaul, in this case from RT 30, the body shown going to RT 20 after overhaul the following month. Note how the driver had adjusted both top and bottom parts of the windscreen to be slightly open to give ventilation on what was clearly a warm day.

Above In addition to war casualties, accident-damaged buses were also apt to disappear off the road for lengthy periods. One instance involved RT 18 which overturned following a side-on collision with a Bedford coach owned by Mills and Seddon of Manchester at the junction of Gunter Grove and Fulham Road in Chelsea while operating on route 14 in September 1945. It did not return to Putney Bridge for service until February 1946.

Below Tucked away in the mews terminus off Baker Street, RT 38 still carries a wartime advertisement and requires its nearside headlamp mask to be removed and new sidelamp lenses to bring its lighting to full strength for the first time. *S.A. Newman*

Most 2RT2s allocated to Victoria garage were recalled for use at the two Putney garages, route 77A exchanging its ST/STL/RT operation for utility G class Guy Arab vehicles on 10th October 1945 (route 52 had been converted to STL operation in October 1943). There had also been another significant departure from the garage earlier in the year when STL 2188 arrived from Hornchurch on 18th August 1945 to displace RT 1, this historic vehicle then being withdrawn from service for the final time. Victoria briefly continued to find work for members of the class, five of which were retained from the 77A conversion for use on route 137. In the event, this allocation was short-lived and RT 48 became the last 2RT2 to leave Gillingham Street on 26th October bound for Putney Bridge.

1946 began with the the entire class based at the two Putney garages apart from RT 97 at Kingston due to its involvement with the PAYB experiments; and RTs 18 and 19 which re-entered service on 1st February and 1st March respectively. The largest number of 2RT2s was to be found at Putney Bridge, where 85 of the type were variously employed on the 80 duties which departed from the garage during the course of each weekday. The remaining 64 vehicles were used to cover the 58 duties on routes operated by Chelverton Road.

The situation was soon altered, the first change occurring in May when the number of weekday duties on route 28 at Chelverton Road increased by seven and, as a result, six vehicles were transferred to the garage from Putney Bridge to assist in meeting the revised requirements. Eight STLs recently ousted from Sutton arrived at Putney Bridge and were allocated to route 14. This seemingly insignificant action was enough to ensure that Putney Bridge would never again be provided with sufficient 2RT2s to meet its total weekday commitments. And here were sown the seeds of an informal 'agreement' between the two garages which decreed that any 2RT2 shortages suffered by Chelverton Road would be restored by transfers of the type from Putney Bridge. The introduction of summer schedules on 22nd May brought 2RT2s to two seasonally extended routes for the first time, the 14A which was an extension of route 14 on summer Saturday afternoons and Sundays to Hampton Court, and the 72A which was a summer Sunday extension of the 72 to Chessington Zoo.

The 2RT2 fleet had ended the war years looking quite shabby, and this persisted for a while as the magnitude of the task of keeping enough over-age and sadly neglected buses of all types serviceable made it impossible to catch up with the backlog of less urgent work for several years. At any one time, although some recently overhauled and repainted examples of the class were to be seen, others on the routes still looked what they were by normal peacetime standards, rather elderly and hard-worked buses. Yet their close resemblance to the much newer post-war fleet made them seem much more up-to-date, and created a perhaps irrational expectation in one's mind that they should look as smart as say the 1946 additions to the STL and STD classes, despite being so much older.

The re-appearance of STLs at Putney Bridge allowed the next round of 2RT2 overhauls to commence (the previous cycle having been completed in the spring of 1945). The first arrivals at Chiswick, during March 1946, were RT 7 from Chelverton Road and RT 39 from Putney Bridge. Both vehicles left the works sporting the same short-lived livery that had adorned RT 19, carrying the prototype body, upon re-entry into revenue service a few weeks earlier. This style had been adopted for repainting overhauled STL buses of most main types for a period in 1946 and for the final batch of 100 Daimler CWA6 buses, D 182-281, new that summer and autumn. In contrast to the livery applied to the 2RT2s when first delivered, this colour scheme was simple, being all-over red with black wings relieved by two bands of cream. The lower band was located in the traditional between-decks position whilst the upper band was in fact the strip of half-round beading which ran above the upper deck windows, just below the roof line. As part of the livery applied to the vehicles when new, most of this beading had been picked out in red and acted to separate the grey of the roofs from the white of the upper deck windows.

Right RT 7 was one of three 2RT2s to be given the short-lived early post-war livery with a narrow cream band above the upper-deck windows. It is seen at Chiswick on 15th April 1946 fresh from overhaul but with damaged valance above the bonnet and with its sidelamps still masked to blackout standards. *LT Museum U37491*

Below Design differences between the 2RT2s and the post-war variants were not only found on the exterior of the vehicles as this view of the lower saloon of RT 7, also in April 1946, demonstrates. Most noticeable is the absence of a bell-cord, two ceiling-mounted bell pushes being fitted, much as on the STL. Also of interest is the black curtain for the front window, this item of soft furnishing later disappearing in favour of the more familiar concertina blind. Less obvious are the green and cream window shrouds which are larger in design, obviating the need for straight sections above and below each unit. It would also appear that the interior of RT 7 is the subject of an experiment as the rails below the front windows have been inverted and there are two additional advertisement panels fitted below them. *LT Museum U37490*

London Transport contributed two buses to the Victory Parade held on 8th June 1946. It would have been more 'democratic' to have paraded an STL and an LT as the two main types in service in London throughout the war, but the chance was taken to show off the RT, basically the standard for the future. So RT 4 and RT 39, in service for almost the whole war period, were selected. Although kitted out with anti-blast netting, headlamp masks and white paint on the front mudguards, they were not painted in authentic wartime livery but in two alternative versions being considered for general adoption. They are seen passing the saluting base in The Mall where King George VI and Queen Elizabeth, Winston Churchill and other dignitaries reviewed the parade. RT 39 is in a style briefly adopted, notably for the last batch of D-class Daimler CWA6 buses, but RT 4 gave a better idea of how most of London's future new buses would look. Both were fitted with front wheel discs, of the type more familiar on the 10T10 coaches, for this occasion. RT 39 is also shown at Chiswick. *LT Museum.*

During April RT 4 became the next 2RT2 to receive attention. It subsequently emerged from the works in May painted in the style of livery that would be adopted as standard for STLs and new and overhauled RTs for the next four years. This suited the RT body particularly well, perhaps not surprisingly as it virtually echoed the original concept seen on RT 1 when it was first revealed to the public – it is often very difficult to better the original form of a true work of art.

On 8th June 1946, a big Victory Parade was held in London and among the participants in this were the recently-overhauled RTs 4 and 39. They were included to commemorate the contribution buses had made during the war, and had wartime-style netting applied (incorrectly) to the outside of the windows and white markings added for the occasion. After the event, RT 39 was soon returned to Putney Bridge garage but another public engagement awaited RT 4. On 7th July 1946, the British Motoring 50th Jubilee Cavalcade took place in the (aptly named) Regent's Park where RT 4, bereft of its netting, was among the parade of 500 vehicles reviewed by the King and Queen.

The First RTs

The overhaul programme then gathered momentum and by the time the year was out nineteen vehicles from Putney Bridge and seventeen from Chelverton Road had been despatched to Chiswick. One notable arrival during June was RT 46 which initially received the body from RT 23 and later the body from RT 94 during a series of nineteen displacements instigated by the arrival of RT 85 at Chiswick on 11th July.

Body transfers between 2RT chassis were far less common than had been the case with earlier generations of London buses. The practice of lifting bodies at overhaul, and switching them to other chassis simply because the body overhaul took longer, had largely ceased during the war and, so far as this class was concerned, was only revived briefly afterwards.

All but 32 of the 150 retained their original bodies right to the end of their days with London Transport, in marked contrast to the pattern before or subsequently, once the intended overhaul procedure for post-war RTs got under way. A few transfers, quite often related to repair of bomb damage, had occurred during the war, and there was a very limited revival of the system of removing bodies for overhaul and fitting them to different chassis during a period from mid-1946 to early 1947, when 19 buses emerged from Chiswick with different bodies from those fitted when they went in.

The enormity of the backlog of body overhaul work which faced London Transport was becoming apparent. Normal methods had been abandoned during the war, partly because of the diversion of much of Chiswick's resources normally devoted to body repair and construction to the London Aircraft Production Group, making bombers for the Royal Air Force. The standard Chiswick-built composite body, most notably as fitted to the STL class buses which formed the largest part of the fleet, was designed for about a ten-year life on the assumption that it would be kept in good order by regular overhaul and repainting during that time. By 1946, the earlier STLs were past that age and it soon became clear that even the later ones had suffered serious deterioration, making anything approaching the normal production flow impossible.

Much London bus overhaul work was put out to be done by outside contractors and, generally, they were not geared to the lifting of double-deck bodies, which needs premises with very generous headroom. In any case this was apt to worsen matters where body structures were badly affected by rotted woodwork, for the lifting process added to the stress on a weakened structure. Thus the whole pattern of overhaul work changed, within Chiswick as elsewhere. Bodywork tended to remain associated with the chassis on which it was mounted, and the 2RT2 fleet reflected this. Most of them are recorded as having received four overhauls during their service lives with London Transport, generally being off the road for between a month and six weeks or so at a time. The intervals between them varied, being around four years in the war and immediate aftermath, reducing to nearer three years in the early 1950s.

Easily recognisable among 2RT2s in the late wartime and early post-war years was RT 108, *above,* another vehicle to have received repair work somewhere other than at Chiswick. The dip in the cantrail moulding at the corners was in line with a fashion trend at the time and was emphasised by the paint treatment. Until the new post-war livery started to appear in 1946, the bus stood out as the only 2RT2 with a red painted front dome. *Bob Burrell.*

Body 309 of RT 41 in the Chiswick paint shop, November 1947, during the short period when bodies were lifted from chassis for their respective overhauls, even though often refitted to their original chassis as in this case. *LT Museum*

A slightly later view, also in the paint shop, this time with bodies travelling through the works undergoing overhaul. The unidentified 2RT2 has been fitted with a vertical route number plate holder.

RT 46 returned to service from overhaul in late August but early in September was back at Chiswick in order to take part in blind display comparison trials with RT 110. Among other matters, the future of the roof number box was under review and RT 110 was an obvious choice, having had its front roof box removed during the war.

Discussions relating to route information to be carried by the post-war RTs had been referred to a committee earlier in the year, its initial recommendation being the deletion of the offside route number, which had tended to fall out of use during the war years. This resulted in the panel immediately to the rear of the last passenger window downstairs, which incorporated a spring-loaded holder, being replaced by a plain panel on overhaul. The early examples of 3RT3s were built lacking the capability of displaying an offside route number until a directive was issued reversing the decision, which also had the eventual effect of restoring this facility to some of the 2RT2s.

Another recommendation of the committee resulted in the whole of the Putney Bridge 2RT2 allocation being experimentally fitted with a device set at an angle of 35° on the front lower deck nearside pillar capable of displaying a number plate measuring $2^3/_4$ inches wide and $9^1/_2$ inches high below the word 'Route'. This innovation, already included in the specification for the 3RT3s, was designed for the convenience of intending passengers who could thereby identify the route numbers of buses when a number of them arrived together at a stop serving more than one route. However, Chelverton Road's vehicles were not so equipped and, when 2RT2s were transferred in with this additional route number facility, it remained unused. Eventually it ceased to be used at Putney Bridge.

The blind layout comparison trials in the autumn were quite a sophisticated affair. Previous decisions not to perpetuate the use of the nearside route number and the rear roof box on the post-war variants had already been taken, their replacements being in the form of revised intermediate destination blinds displaying route numbers that effectively reduced the number of via points shown. Presumably, in order to assist the decision making process, photographs of RT 1, when new, were made available in which the front and rear roof boxes were professionally 'touched out'. Earlier in the year, widening of the front roof box had been considered, but now its use was to be compared with alternative methods of showing the route number.

In addition to its absent roof box, RT 110 had undergone further surgery, during its 1946 overhaul, in order to complement the revised front blind layout which it was to display. This involved the removal of the vehicle's canopy valance and the installation of a short length of flat beading to replace a half round piece of similar length which ran immediately below the front windows and for a short distance on each side of the vehicle. The second modification was to enable retaining clips to be fitted – four at the front and two at the side – which

would secure panels displaying alternative 'sans roof box' blind layouts. Years later the four front clips were still in evidence on the bus. To complete the presentation, RT 110 was also equipped with new style side and rear intermediate blinds as described previously, and a dummy route number blind assembly below and to the nearside of the canopy. On both this vehicle and on RT 46 the ultimate destination display was moved from its position above the intermediate blind to below, an arrangement which would enable winding from inside the driver's cab.

The blind display experiments carried out on RTs 46 and 110 at Chiswick in the late summer of 1946 also included a review of the various metal route number plates then still in use. A new feature for the RT2s, and the first RT3s, was a vertical plate under the front canopy, illustrated far left and below, the result of continuing concern about the best way to incorporate a route number which would not be obscured when a number of buses were at a bus stop at the same time. This same concern is evident in the two front end photographs of RT 110.

RT 46 was also provided with a section of flat beading but only two retaining clips, the demonstration forward blind display being merely an inverted form of its own. Old style side and rear blinds were installed and both roof boxes brought into use – this being the very last occasion when a rear roof box was used to officially display a London Transport route number.

After examining the 2RT2s, the committee found in favour of the grouped destination and route number display shown on RT 110, the repercussions of this decision sounding the death knell for the roof route number box in London.

1947 was to be the year of the 3RT3 and, in anticipation of these prestigious vehicles entering service, arrangements began to be put in hand for drivers at garages due to be amongst the first recipients to be type-trained. The 2RT2s were an obvious choice and the year was only a few weeks old when RT 38 was transferred to Leyton on 17th February to undertake training duties. The staff at the east London garage appear to have been anxious to try out their newly acquired steed under full service conditions and photographs exist showing RT 38 working on routes operated by the garage, the vehicle's full PSV licence making this possible.

RT 38 remained at Leyton until September when it was transferred to undertake further training duties at Potters Bar, Turnham Green, Catford and Old Kent Road; not returning 'home' to Putney Bridge until February 1949. Its duties at Leyton were immediately taken up by RT 56 which later moved on to Middle Row, Holloway and Mortlake, these secondments lasting about twelve months in total. RT 29 and RT 50 were the only other members of the class to be used in a similar capacity during the year, RT 29 going to Bromley in November and returning to Putney Bridge in June 1948 via Seven Kings. But the record for these early migrants must go to RT 50 which departed Putney Bridge for Middle Row in December and later toured Holloway, Tottenham and Forest Gate garages, before arriving at Chelverton Road in July 1950!

Fifty-five 2RT2s had their second overhauls started during the course of 1947, a year which also saw RT 85's return to

passenger service in March – its chassis having been stored at Chiswick since the previous August. Another notable departure from Chiswick in September was RT 121 which left the works fitted with quarter-drop windows, followed by RT 101 in November with the same alteration. In all six 2RT2s took part in the experiment which led to the adoption of this feature on the RF; the remaining four were RT 127, RT 129 and RT 84 which re-entered service in January, April and June 1948 respectively and RT 82 in February 1949.

The front upper-deck winding windows were shallower than those on the sides. It was rare to see front windows on any double-deckers opened more than a few inches because the upper deck became excessively draughty otherwise. As part of the experiment therefore an opening of just three inches was provided. All six buses retained the shallower opening windows until withdrawal.

2RT2 operation on route 72A proved to be brief and lasted just two seasons. The route was transferred along with the 72 to Hammersmith garage on 12th November 1947. As part of the same programme of changes, Battersea's share of route 28 was returned to Chelverton Road and officially saw 2RT2s for the first time since the end of July 1940.

January 1948 brought an allocation of post-war RTs to Middle Row garage to replace STL duties on routes 15 and 28. This resulted in the first instance of 2RT2s and their post-war counterparts sharing the entire operations of a route, and we are grateful to Gavin Martin, an engineering-based historian who spent periods with both London Transport at Chiswick and AEC during the earlier parts of his career, for contributing a reminiscence of travelling as a passenger on the route between West Hampstead and Golders Green from time to time during this period. He recalls his surprise at finding that the advantage in this comparison, particularly in terms of performance on the road, lay decisively with the older buses. Broadly speaking, the increased weight of the new buses, 7tons 10cwt as against 6tons 15cwt, was compensated by their increased power, 115bhp compared to 100bhp.

It was evidently a combination of several factors, one of them human, in that the Chelverton Road drivers invariably started from rest in first gear whereas the Middle Row drivers were trained to follow London Transport's post-war preferred practice of starting in second. Rapid response at lower speed may also have resulted from the 2RT's lower gearing, with 5.75 to 1 rear axle ratio as against 5.166 to 1, but another factor was the reduced inertia of the 16in diameter fluid flywheel on the older buses as compared to the post-war 18in version with cast steel backplate.

In terms of smoothness of gear changes, the 2RT gearboxes again came out better than those of the 3RT buses, which at that stage used the same band setting for all forward indirect gears, which meant that third gear was unnecessarily fierce in taking up the drive, the combined effect of this and the heavy flywheel resulting in jerky gearchanges more often than not.

RT 129, one of the six 2RT2s fitted experimentally with quarter-drop windows in 1947-49, is seen a couple of years after its 1948 overhaul a little the worse for wear. It is one of a number of 2RT2s given brown roofs about this time, though most were overhauled with red ones after the war. At overhaul it has lost its offside route number plate holder in accordance with the early post-war policy.
Alan B. Cross

Other recollections of the relative performance of the 2RT and 3RT buses vary, and clearly the mechanical condition of individual buses as well as techniques employed by drivers had an influence. A common problem with all fluid flywheels at that period was a tendency to leak oil from the output shaft gland, just in front of the drive coupling to the gearbox. Even when slight, the consequent drop of oil level within meant that the normal responsive take-up of drive gradually became less effective, with an effect akin to clutch slip on a vehicle with conventional gearbox. In fact the drivers' reports when this became noticeable would describe the fault as 'slip'. Alan Pearce, then a mechanic working on 2RT2s at Putney Bridge garage, recalls that the glands needed frequent attention.

In basic principle the gland was not unlike that on a domestic tap, but could need tightening as frequently as every ten days or so. It was then generally replaced at the 24-week rota docking, and wear on the flywheel bearings, which also had an influence on the problem, would normally cause replacement of the complete unit with an overhauled one at the 48-week rota interval.

It may be that this problem had caused the Chelverton Road drivers on route 28 to regularly use first gear for starting from rest. If then applied to a bus with everything in tip-top condition, good acceleration undoubtedly resulted. The 2RT2 was certainly a "lively beast", Alan Pearce recalls, "but with a 'lazy' driver and a fluid flywheel in need of attention, it was quite pitiful to hear a '2' getting to grips with a full standing load".

A major change to their mechanical specification was made in the period around 1949, this being the conversion of the engines to the toroidal form of direct injection from the original pot cavity system – Alan Pearce reports the work to have been completed on Putney Bridge's allocation by July 1950 and he is confident that this would also have been so at Chelverton Road by around that date, because frequent exchanges of vehicles took place and any unmodified vehicles would be immediately identifiable. Unfortunately the vehicle cards did not record the alteration, it being carried out during engine overhaul.

The change would have involved considerable expenditure on new parts, including pistons, cylinder heads and injectors plus many smaller items, and implies an expectation of quite lengthy continued service at the time. The effect was immediately obvious to the bystander, quite apart from passengers or the driver, for the distinctive quite deep engine note, which had enabled the interested pedestrian to identify a 2RT in original form even when approaching from behind, gave way to the lighter but slightly harsher tone hitherto characteristic only of the 3RT.

Alan Pearce's view is that the related change of the 5.75 to 1 rear axle ratio to 5.2 to 1, as applied with the original toroidal conversion of RT 19's engine in 1940, was not made as part of the later conversions of the class as a whole. However he feels almost sure that RT 52, which was often used for mechanical experiments, did receive a 5.2 to 1 differential unit at some time. RT 52's chassis had spent some time at AEC in 1944-5 before that of RT 19 was chosen to become the prototype for the post-war fleet. It spent lengthy periods on routes 85 and 93, probably to carry out fuel consumption tests involving Wimbledon and Kingston Hills. However, the record cards show that most buses were converted to 5.2 to 1 eventually, possibly when worm drive units required renewal or at chassis overhaul.

The toroidal conversion would have made it quite possible to increase the power output to the 115bhp of the standard 3RT but in view of the lighter weight, and especially with the unchanged axle ratio. it seems quite likely that the engines were derated back to somewhere near the original 100bhp, quite simply done by adjusting the injection pump output. Even so, the engines would have different characteristics; probably pulling slightly better at low speeds, all else being equal.

RT 39 received four different colour schemes during its time with London Transport. Here the vehicle is seen in its penultimate livery working on the summer Saturday afternoon and Sunday route 14A. *D.A. Jones*

With the increase in the number of garages receiving post-war RTs came an increase in the demand for suitable training vehicles. In addition to the 2RT2s already serving in this capacity six additional members of the class were loaned out from January to March 1948, all of which were allocated to Putney Bridge with the exception of RT 7 which was supplied by Chelverton Road on 25th March. No further vehicles became trainers until May when the arrival of post-war RTs at Putney Bridge released six on secondment; three of which went to Dalston. The second half of the year witnessed an additional nine 2RT2s moving to temporary work on training duties.

The arrival of post-war RTs at Putney Bridge resulted in changes to the garage's allocation of vehicles to routes. The eighteen STL duties on route 14 were withdrawn and the vehicles reassigned to routes 85 and 93 where a total of 21 duties were covered by this type. Route 14 then became wholly RT operated using both 2RT2s and post-war RTs – the second time such an event had occurred during the year but this time with both variants operating from the same garage.

A total of 47 2RT2s were identified for overhaul during 1948, among them RTs 73, 74 and 128 which had been selected as the first of the type to be uprated to a new specification upon their entry into Chiswick in July. The change to a reciprocating compressor, with this and the dynamo mounted on opposite sides of the gearbox and belt driven from it, as fitted to RT 19 in 1945, had been adopted as a general modification for the remaining 99 2RT2 buses with the original shaft drive for these auxiliaries, these being recoded 3/2RT2/2. The alterations to the body centred mainly on the provision of an additional floor trap and the realignment of the forward leg of an adjacent seat frame in order to facilitate access. It took until October 1951 for the programme to be completed by which time one of the vehicles planned for this conversion, RT 85, was destroyed by fire before the work could be carried out and the components that were to have been used were applied to RT 52. This was the only example of conversion to this specification from the interim 1/2RT2/1 form, which had retained the shaft driven dynamo, and the last to be completed.

1948 was the year of the London Olympics, the first Games to be held after the cessation of hostilities, and 2RT2s were among the varied types used to provide transport to the events. One such vehicle was RT 138 which was then currently on loan to Catford garage as a trainer, in which capacity it would move on to Elmers End and Sidcup before returning to Chelverton Road in January 1949. *LT Museum*

20 Safe As Fire Destroys London Bus

Thirty-six 2RT2s were seconded as driver trainers during 1949, four of those involved being absent from official service activities for the whole of the calendar year. Of the remainder, careful management ensured that no more than half were away from their parent garages at the same time, an absence which in some cases was extremely brief; none more so than RT 32's transfer to Battersea in November which lasted all of three days! Large scale use of 2RT2s on training duties for short or extended periods was to continue.

During the year, 29 2RT2s entered Chiswick to be over-hauled and converted to the 3/2RT2/2 specification. This programme ran alongside the overhauling of those members of the class converted as 1/2RT2/1 of which seven out of the 50 members of this sub-group also entered the Works.

The spectacular end of one member of the class occurred on 14th May 1949, when RT 85 burnt out in the Cromwell Road whilst in service on route 74 from Putney Bridge. The sequence of events that led to the conflagration started from the failure of a pin in the accelerator linkage, causing the engine to race away and leaving the driver with no means of regulating its speed. In an attempt to control the situation the vehicle was put into gear and the fuel tap was turned off but this action failed and brought about the collapse of the fluid flywheel. Hot oil then ignited under the floor and the fire took hold to such an extent that the vehicle was burnt out. The remains were later towed to Aldenham Works where they were dismantled the following month; the bus having been officially withdrawn on 21st June. As a direct result of the incident, a warning notice was placed in all buses which gave directions to drivers regarding the actions required to stop engines in an emergency, the first being to select neutral gear. A redesign of the RT accelerator linkage followed which involved fitting a spring to the fuel pump lever, this modification being applied to both new models and the existing fleet.

Firemen had to keep at a distance because of the heat when they fought the blaze which reduced the bus to a blackened frame work

The burnt out bus.

"Star" Reporter

A LONDON bus was burnt out in Cromwell-road, Kensington, today.

The 20 passengers in it escaped.

Smoke was seen coming from the bonnet of the bus, a double-decker on Route 74, going from Camden Town to Putney.

The driver pulled up outside the Majestic Hotel and, with the conductor, investigated.

One of the passengers, Sister Clare Herbert, of the Naresborough Nursing Home, Kensington, said: "The driver had been having some trouble before. Somebody came over and threw a bucket of water over the bonnet.

"Then, suddenly, the flames burst out, and soon the whole bus was alight. I ran to the nearest telephone and dialled 999."

Crowds collected and traffic was held up as three fire engines arrived and their crews fought the blaze.

Mr T. Pascoe, manager of the Majestic Hotel, said: "The whole bus was burnt out in less than 20 minutes. When the fire was at its peak the flames were rising above the level of the roof of our hotel, which is a five-storey building. The heat was intense."

A collapsed flywheel was the cause of this conflagration which resulted in RT 85 becoming the first 2RT2 vehicle to be withdrawn. The London Evening Star made the incident its main front page story. After the fire, which occurred whilst the vehicle was in service on route 74 on 14th May 1949, the remains were taken to Aldenham and were later dismantled. The body carried at the time of destruction was that originally fitted to RT 56.

Right Loans of the 2RT2 vehicles are not recorded in the official records of London Transport and one such example involves RT 2. In this view the vehicle is shown operating on route 53 from Plumstead garage on 27th August 1949. *Alan B. Cross*

Below left Loaned to Leyton for driver familiarisation duties, RT 20 was placed in service on routes operated by that garage. Here it is shown on route 10 at Victoria in September 1949. *Alan B. Cross*

Below right RT 19 looking rather shabby at Aldwych in 1950 – eleven years after RT 1 with the same body was shown to the press here. In this view, a vertical bar has been placed in the platform window, this curious addition disappearing after the vehicle's first overhaul, to be replaced by a similar fitment in the rear emergency exit. Of the non-standard items provided only on the prototype, the offset roof ventilator still remains together with the two adjacent opening windows on the nearside of the upper deck. *Alan B. Cross*

After it had received the body of RT 1, RT 19 remained a test bed vehicle for most of its life. There were fairly frequent visits to it at Putney Bridge by engineers from Chiswick, using the bus to evaluate such things as gearbox oils during test runs over hilly roads.

The platform covering on this bus was changed a number of times during the late-1940s. Standard wood slatting gave way to a hard speckled compound and, later, a substance that had the appearance of cork replaced this except around the edges. About 1950 a flat hard rubber coating appeared with a pattern not dissimilar to tyre tread. Finally it was given an all-over covering of knobbly black rubber, which gradually became standard on all RT family buses from 1951. A bell cord was also fitted in the lower saloon of RT 19 after the war.

The fareboard holder on the 2RT2s was on the platform, facing the entrance. Probably before the delivery of the RT3s, RT 19 was fitted with an additional one in what was to become the RT3 position. Later it was also fitted with a horn push to the left of the driver, as on RT3s, retaining also the bulb horn on the right.

During 1949/50 the driver's seat on RT 19 was changed a number of times to evaluate different designs. Ironically, immediately beforehand RT 19 had the seat originally fitted to RT 1, which was more comfortable than the revised design fitted to the 2RT2s from new. On these, with the seat back rather tall and upright and the cushion virtually flat, the driver tended to be pushed forward from the lower spine area. This caused many of the drivers to suffer backache, and London Transport's Chief Medical Officer made several visits to the garages to investigate complaints made on the drivers' behalf by the Transport & General Workers Union.

By 1950 the cab seat design that was later to become accepted was fitted in RT 19. The design was subsequently fitted to all post-war RT family buses on overhaul, but 2RT2s only received the new seats when transferred to training duties. The reason for this was that it was necessary to give good access to the cab for the instructor. As had been 'custom and practice' in the driving school for years, the glass in the window behind the driver was removed from training buses to enable communication to be made between instructor and his pupil. That done, the instructor had the additional facility, in an emergency, to reach forward into the cab and grab the handbrake lever. On 2RT2s fitted with the original seat, this manouevre was almost impossible. Training vehicles with the modified seat retained it if returned to passenger service later, but other vehicles continued to cause drivers some discomfort to the end of service work. Trainers used on the skid pan at Chiswick were at some point fitted with seat backs about nine inches high only, giving maximum access from the saloon.

At the garage RT 19 was treated by the engineering staff as a 3RT, as all chassis components in its later years were as on a standard 3RT with the exception of the air system, which had features that were unique to this bus.

The vehicle was the subject of a complete overhaul which lasted from 7th March to 1st May 1951 during which time the body was fitted with standard mudguards, repainted in the new all-red livery and traded its unique six louvre front panel for a more conventional seven louvre example. It returned to Putney Bridge until final withdrawal from public service on 1st November 1953. At first thought, it seems rather wasteful that what should have been a good bus, with chassis to post-war specification and a five-year Certificate of Fitness granted just after the 1951 overhaul and thus with 2½ years to run, should have been taken off the road. However, London Transport's stock position had greatly improved by then.

Another kick to the mystery of RT 19's correct coding is made by a note made by Alan B. Cross in January 1953 which records the body plate on the bus as then reading RT1/1. So there was still a discrepancy between what records and the bus itself conveyed. What the significance of the '/1' might have been is a matter for speculation – there had been the reversion to 56-seat capacity in 1946, or another possibility was the fitting of standard mudguards. The plate still read thus in 1956.

RT 19 became something of a guinea pig after receiving the body from RT 1. Nothing indicates this from the outside in this view; almost all the experiments undertaken with the vehicle being on the engine and chassis and on the body interior (see text). This view illustrates well the slight inward curve of the waistband above the driver's cab, where a ventilator intake was fitted. The driver's mirror is of a type fitted to a number of buses in the early post-war period, one of similar appearance being fitted to the prototype RMs. This photograph was taken after the vehicle's last overhaul in 1951.

Vehicles from other classes continued to be despatched to Putney Bridge in order to augment its dwindling 2RT2 allocation. In May 1949 six new Bristol K5Gs then on loan from Tilling companies were received and put on route 93, whilst the input of brand new RTs 2236-2243 was another measure.

During 1950, the number of 2RT2 overhauls being handled at Chiswick reached its peak. In the space of twelve months, 37 unmodified 2RT2s were overhauled and uprated to the 3/2RT2/2 specification, whilst 37 of the 1/2RT2/1 type also received attention as part of the existing three-year works cycle. The length of time for overhaul ranged between six and eight weeks with any deficiencies in allocations supplemented by transferring STLs into Putney Bridge, the garage still being required to make good any shortfall in Chelverton Road's unique 2RT2 stronghold. Eventually, staff at Putney Bridge rebelled as they felt the situation had been allowed to continue for far too long and, as a result, STLs began to be sent directly to Chelverton Road to act as temporary replacements for missing 2RT2s. Some months later, Chelverton Road staff also voiced their displeasure and received a response that improvements could be expected during November.

3rd May 1950 saw the introduction of RTWs on route 85 from Putney Bridge to replace the Monday to Saturday STL and Sunday RT allocation: although, as part of the change, four Saturday duties were worked by RTs until October. Eleven days later, on 14th May, the RTWs were transferred out and replaced with a temporary allocation of RT family members. This action was necessary in order to release the RTWs in the first of three trials involving the 8ft wide buses across central London as part of a large scale experiment to ascertain whether they could be used in busy and relatively narrow streets without worsening traffic congestion.

The initial trial, which ran from 15th to 19th May, was based on routes passing through Notting Hill Gate and amongst the garages that participated was Chelverton Road, whose allocation of 2RT2s on route 28 was exchanged for RTWs. The temporarily displaced 2RT2s were then sent to Putney Bridge to replace post-war RTs for use at Upton Park where they liberated some RTWs for use on Merton's part-operation of the 88 – another route involved in the first trial. The second trial, from 19th June to 2nd July, was based on routes serving Shaftesbury Avenue and therefore included route 14 operated by Putney Bridge and Holloway garages. However the final trial, concentrated on Threadneedle Street in the City during July, did not require the participation of either Putney garage.

Route 74 from Putney Bridge underwent a number of changes in 1950. The number of weekday duties operated by the garage had increased to 28 in January and, until 3rd May, was operated by a mix of post-war RTs and 2RT2s. On that date the allocation changed with STLs taking up half the number of duties, a situation that was to remain until September when RTWs assumed full control.

Above RT 35, overhauled with a brown roof at Chiswick in November 1949, is seen at Morden in January 1950. Like most of the class, it retained its original body throughout its life. At its recent overhaul it had acquired the belt drive for compressor and dynamo, becoming 3/2RT2/2 in consequence. *Alan B. Cross*

Below RT 134 is photographed on 6th October 1949 operating from Chelverton Road garage on route 30. The front roof box is one of a few with its glass painted black; red was normally used for masking. No member of the 2RT2 class was ever equipped with semaphore arms or flashing direction indicators whilst owned by London Transport. *Alan B. Cross*

London Transport decided to simplify its livery in 1950, vehicles beginning to appear in almost all-red, with only the inter-decks stripe in cream, from April of that year. Most observers seem to agree that it was a less attractive style, though the quality of finish remained good, with London Transport's elegant types of lettering and the retention of the traditional black mudguards helping the overall effect. The 2RT2s shared in the change, the process for these vehicles being completed by 1953. Although the modified livery did nothing to improve their appearance, it is fair to say that during the early 1950s the overall look of the batch did generally improve as the result of a period of more normal overhaul cycles. For a period prior to the change, a few overhauled 2RT2s were delivered from Chiswick with roofs painted in a shade of brown similar to the treatment on utility buses and STLs at the time.

RT 84 shows the all-red livery in Upper Richmond Road. The vehicle is equipped with quarter drop windows but, despite this modification, no additions were made to its RT2 and, later, RT2/2 body codes to show this modification.

Two 2RT2s at Golders Green in 1952, of which RT 9 is fresh from overhaul. Although Chiswick had started to restore full blind displays to overhauled vehicles from about the middle of 1951, 2RT2s continued at this time with the restricted style. *F G Reynolds.*

RT 22 in process of recovery following its accident on Wimbledon Hill on the second day of 1951 whilst working on route 93 from Putney Bridge garage.

Quick braking by the driver of RT 100 on 17th May 1951 prevented injury and greater damage when a large elm tree fell in its path at Putney Heath while operating on route 85 to Kingston. The 15 passengers were unhurt.

The atrocious weather conditions which prevailed during the winter of 1950-51 resulted in a second 2RT2 being written off. The vehicle involved was RT 22 which overturned on Wimbledon Hill in icy conditions on 2nd January whilst operating on route 93. The unfortunate vehicle then languished at Chiswick until 4th May on which date the body was removed and scrapped, the chassis being dismantled on 20th June.

The STLs at Putney Bridge officially departed on 5th May when the garage's share of the 93 was given over to complete RT operation for the first time in three years. However Sutton's contribution to the route comprised Ds and STLs and this was destined to last for another eighteen months.

Many early RTs had transferred to the training fleet during 1950 as intensive preparation began to train tram drivers on buses. In February, RT 65 became the first 2RT2 to transfer to the base established especially for this purpose at Camberwell garage; followed by RT 101 in March. The 2RT2 presence during the ensuing months was minimal with no more than three vehicles at Camberwell at any one time. An increase occurred in July 1951 when RT 43 joined RTs 24 and 132, the number doubling in August with the arrival of RTs 10, 35 and 94 from Putney Bridge. Walworth garage acted as a temporary home for the trainers from October 1951 to January 1952 where RT 124 replaced RT 132. On their return to Camberwell, RTs 5 and 40 replaced RTs 10 and 24; those that remained were licensed for service at Peckham on 7th July.

Photographed in June 1952 whilst on loan for a few months to Reigate garage, RT 124 is seen in Redhill on a short working of route 405. This vehicle also served on routes 410 and 414 from the same garage although it is interesting to note that its transfer to the Country Area went unrecorded, the bus being officially allocated to Camberwell on training duties during this period. *Alan B. Cross*

A transfer of duties in December 1952 broke the 100% operation of 2RT2 buses by Chelverton Road that extended back to 1941. Ten workings on the 74 transferred from Putney Bridge brought the 8ft-wide Leyland RTW type into the garage's stock. Here RTW 391 is seen with RT 80, another bus that had run from Putney Bridge for much of its life. *Lens of Sutton*

RT 31 stands outside Dorking garage having just completed a journey on the Sunday extension of route 93 from Epsom, which ran during the summer for most of the time 2RT2s operated on the route. *Alan B. Cross*

RT 95 pauses in Wandsworth High Street on a short working of route 37 which will take the vehicle as far as Richmond. The red masking on the inside edges of the front blind box glass is a remnant from the period of restricted blind displays.

In an operational sense, the concentration of the 2RT2s on a limited number of routes and garages, mostly not changing over long periods, began to alter. The conversion of the London tram system, which had largely been concentrated north of the Thames in the mid to late 1930s, using trolleybuses, was resumed in the autumn of 1950, this time concentrated in south London and using motor buses. The early stages of the conversions were based on new vehicles, impressive-looking fleets of shiny RT and RTL buses appearing on the streets in place of the old trams, but in July 1952, at the final stage of the process, a more mixed fleet was used, including 2RT2 buses for the first time. As the final echoes of Auld Lang Syne died away in the early hours of Sunday 6th July, buses, drafted in for use on routes replacing the last of the capital's tram system, were being prepared for service. In addition to the five 2RT2s from Camberwell, a further thirteen members of the class arrived at Peckham, most having been released from training duties at south London garages. The vehicles involved were RTs 5, 11, 14, 17, 21, 35, 40, 43, 51, 64, 69, 71, 94, 98, 121, 124, 127 and 140.

Accompanied by an STL on the 163, which like the 177 was a tram replacement service, RT 98 is seen on the tramway reservation along the Victoria Embankment. Later in life, RT 98 was to be the star performer on the Chiswick skid pan. *D A Jones*

Despite having officially become a bus garage, with an allocation of 79 vehicles, on the day it despatched London's last tram to the scrapyard, New Cross was in no fit state to undergo such a swift change, and Peckham had therefore been selected as one of three garages to provide temporary accommodation for its fleet. Thus the 2RT2s, fifteen new RTs and a mixed bag of STL types took to the streets on tramway replacement services 163, 177 and 182 upon which duties they were required to display New Cross plates (code NX). This practice is known to have occasionally lapsed with Peckham's short-term tenants being fitted with PM plates when employed on the garage's own routes, such as the 70A.

The arrival of RT 117 at Peckham, from overhaul, on 8th October resulted in nineteen 2RT2s transferring to New Cross when the garage re-opened on 22nd October where they were joined, on the same day, by RT 20 from Chelverton Road. However, the actual number available was reduced by two – RTs 94 and 35 both undergoing overhaul on the day the converted tram depot re-opened its doors. With three garages now operating 2RT2s, the overhauling and transfer system became a little more complex. This resulted in 33 2RT2s seeing service at New Cross, at different times during the 1952-1955 period, with a peak being reached in mid-1953 when 24 were available for passenger carrying duties. Members of the class new to the garage soon began to arrive straight from overhaul, whilst some, which had formed the initial batch of twenty, departed for periods of further service at Chelverton Road and Putney Bridge or to take up training duties. One notable new entrant, during October 1954, was RT 62 which later became the only 2RT2 to operate from the five main garages with which the class was associated. Upon RT 62's withdrawal from New Cross routes on 16th March 1955, RT 151 arrived as replacement and remained in revenue service for just nine days, its PSV licence being surrendered on 25th March.

The tramway replacement routes from New Cross were included in intermediate blind experiments which involved just three via points being shown. RT 5's participation on the routes was exceptionally brief, the vehicle entering Chiswick for its final overhaul, lasting five weeks, during its three and a half month secondment to Peckham in 1952. Upon transfer to New Cross in October, the bus saw just four months in service and a similar length of time as a trainer before entering a final four month period of passenger carrying duties in July 1953 from Chelverton Road. RT 10 *far left* was unusual in combining the old livery with full blinds and is seen at the London end of the 177. *Left* 2RT2s appeared on routes 70 and 70A on two separate occasions; first between July and October 1952 when the type was temporarily allocated to Peckham, the second when the routes were transferred from Peckham to New Cross in January 1954.

CHAPTER SEVEN

The Last Years in Service

Below RT 33 was temporarily transferred to Sutton garage as part of the special arrangements for the Coronation in June 1953. 2RT2 types were in regular use on route 93 from Putney Bridge garage, but Sutton garage, which shared the route, normally provided vehicles from its allocation of RTLs. *Alan B. Cross*

Below right Loaned to Hendon garage for the period of the Coronation, RT 27 is shown sporting a rosette in celebration. *Michael Rooum*

London Transport's engineers responsible for the specification of the 2RT2s placed in service in 1940 would have expected them to be withdrawn somewhere around 1951 in what then seemed the normal turn of events, and would have assumed that London Transport's first-class overhaul standards were kept up over that period. So the survival of most in passenger service until 1955 could be regarded a creditable achievement in view of the unavoidable wartime neglect. Indeed a 15-year life was the post-war aim, the revisions to the RT body design made before production had resumed in 1947 being largely addressed to that increased life.

The 2RT2s were not of course alone in having their working lives extended by the austerity years of post-war Britain. At the time of the Coronation of Queen Elizabeth II in June 1953, when final 2RT2 withdrawal from Central Area service was almost exactly two years away, the last of the pre-war STLs still soldiered on beyond their intended life-spans. The 2RT2s however were still being overhauled, such work continuing until the end of 1954. This would indicate an anticipated end for the bulk of the 2RT2 fleet around 1957 if the vehicle surplus and consequent withdrawal of non-standard types in 1955 had not occurred.

All overhauls of the 2RT2s continued to be undertaken by Chiswick, none being consigned to the then-developing Aldenham Works. Eventually the overhaul programme was wound down and RT 135 became the last of the type to receive a full overhaul, emerging from the works on 16th December 1954, two months after its arrival. Although RT 103 entered the works for a fifth time, a week later than RT 135, its overhaul was not completed and the bus failed to re-enter revenue service, eventually being relicensed as a private trainer in August 1955.

RT 66 was easily recognisable until the end by having its offside route number holder set further back in approximately the same position that would identify the post-war RT bodies built by Saunders Engineering. RT 66's original body had been destroyed in one of Germany's flying bomb attacks in June 1944, the body being scrapped at Chiswick the following month. Shortly after, the body of RT 87 became available, itself having been repaired following blast damage from one of the earliest flying bomb explosions. It seems likely that the non-standard staircase panel was installed at this time.
2RT2 Bus Preservation Group, J.C. Gillham

Overhauls on the 2RT2s, generally undertaken at 3-yearly intervals, were not as extensive as those carried out on the post-war RTs at Aldenham in later years, particularly in respect of the bodywork. Colin Curtis recalls that Chiswick was asked to keep costs down on 2RT2 overhauls when the vehicles were about ten years old. While the mechanical side was kept in good order, body overhaul was principally concerned with maintaining a reasonable level of appearance, necessary for each bus to gain a renewed Certificate of Fitness. Retrimming of all the upholstery, for example, was not undertaken, seat fillings and moquette being replaced as and when required rather than at fixed intervals. Some vehicles were to be seen late in life with their original moquette design and a few had a mixture of old and new. All Chiswick overhauls involved a full repaint inside and out, however. Ironically, more money was spent on the last overhauls from about 1953 – hence the long lives some of the buses had with subsequent owners.

The first 2RT2s to receive full blind displays since the end of the war were those that transferred to south-east London to take up duties on the tramway replacement routes in 1952; the two Putney garages had to wait until the latter part of 1953 before a start was made to replace their austerity displays. Eight were noted as having been completed by the end of the year in a notebook kept by photographer Gerald Mead who lived in Richmond and worked in Putney at the time. He also recorded that the majority of the buses received full blind displays between June 1954 and January 1955 – the last of the restricted layouts disappearing from 2RT2s at Putney Bridge and Chelverton Road in March 1955, just two months before the type was withdrawn from Central Area service. The rear roof boxes continued to remain out of use, even when full blind displays were reintroduced, although RT 129 ran for a time with a post-war RT 'AA' canopy number blind in this position.

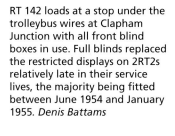

Facing page RT 95 from Chelverton Road garage passes Holloway's RT 2232 at King's Cross in 1953. Holloway's post-war RTs shared the 14 with pre-war RTs from Putney Bridge garage, the post-war variety being the more numerous on the route. The back of the speedometer unit on RT 95 can be glimpsed in the 'droop' of the lower cab windscreen. This unit, which was almost identical to that fitted to the post-war RTs, was a later addition replacing the STL style speedometer housings when cab heaters and windscreen demisters were fitted to the class. *John Fozard*

Left Richmond High Street provides the background to this view of RT 24 on route 37. According to official records, the vehicle was only operated by Chelverton Road garage during the whole of its time in revenue service although it spent a year at Camberwell and Walworth garages in order to train tram drivers in the noble art of driving a bus. *John Fozard*

RT 142 loads at a stop under the trolleybus wires at Clapham Junction with all front blind boxes in use. Full blinds replaced the restricted displays on 2RT2s relatively late in their service lives, the majority being fitted between June 1954 and January 1955. *Denis Battams*

RTs 25 and 100 both received modifications to their upper deck front windows which gave a slight peak effect similar to that on the prototype body. RT 100, at Baker Street, appears either to have broken down or been involved in a minor scrape. Its driver is having a discussion around the back. The detail view is of RT 25.
A.M. Wright, Michael Rooum

With the missing front blind box marked by a replacement panel and two rows of screws, RT 110 was certainly the odd man out of the 2RT2s whilst in this condition. In 1954 it received a replacement front roof box of slightly different appearance from the standard, the glazed area having square corners. RT 19's roof box was similarly modified in 1951 (see photos on pages 91 and 119). *Gerald Mead*

Bottom right Easily mistaken for a post-war RT, in this view, with only the inward curvature of the bottom compass panel and the registration number giving the game away, RT 15 is shown operating on route 70 in Greenwich on 12th February 1955. The vehicle's rear roof box was removed at the time of its 1954 overhaul. *Gerald Mead*

Far right During the rebuilding of New Cross garage, route 163 was spread between Camberwell, Rye Lane and Peckham garages with the latter establishment sharing the service on the route only at weekends using four RTs. Upon its reopening, New Cross provided RT operations on Sundays only with full daily responsibility from May 1958.

RT 73, seen approaching Hyde Park Corner bound for its Putney home, was to earn a place in history after sale in 1959 when a group of students took it to Moscow and back. *R.F. Mack*

By the beginning of 1955, it was clear that London Transport had more buses than it needed. 1950 had proved to be the peak year for bus travel and, although the number of vehicles scheduled for service had continued to rise slightly for a further two or three years, from 1953/54 orders for new RT-family buses were scaled down. Even so, when production stopped in 1954, there proved to be a surplus, and 144 new buses were put into storage. Not only had the last of the pre-war STL and the wartime utility buses been taken out of service, but inroads were being made into the early post-war types such as the final STL buses on crash-gearbox AEC Regent II chassis, the Park Royal bodied D-class Daimler CWA6 batch and the Leyland Titan PD1 buses that had been added to the STD class, all dating from 1946.

Logic pointed to the 2RT2 fleet as the next to go, despite its generally good order and the investment that had been made in conversion of the engines to toroidal. Withdrawals began in February 1955 when new schedules cut the total number of buses needed for service in London Transport's Central Area by 101, with 72 more the following month. Sixteen 2RT2s were withdrawn from passenger service in February and 45 in March as a result of these cuts. At Putney Bridge, RTW buses were brought in from various garages to work on route 14 for the first time since the early 8ft-wide vehicle trials, and at Chelverton Road post-war RT buses took over. A few months later both garages were to receive RTL buses in a general re-allocation designed to standardise on Leyland-built buses in west London garages.

A number of passengers aboard RT 45 have spotted the photographer who is taking this splendid broadside shot of the vehicle at Clapham Common on 27th May 1955. Four days later, on 31st May, this vehicle was withdrawn from passenger service along with the remaining members of the class on Central Area routes. *John C. Gillham*

The end for the 2RT2 so far as passenger operation in the Central Area was concerned came on 31st May 1955. On that final day there were 33 examples available for service from Chelverton Road and 16 at New Cross, Putney Bridge having already completed its changeover following the withdrawal of its remaining 29 licensed 2RT2s on 16th March.

Most of the vehicles remained in London Transport ownership, and indeed the number of the type recorded as owned at the end of each year dropped only slightly from 146 at the end of 1954 to 127 for 1955, at which figure it was to remain constant until 1958. Most were kept on the strength for driver training or staff bus duties. Only 18 were regarded as surplus to requirements and sold off in December, to which number one more, RT 106, could be added in the sense that it was

transferred to the service vehicle fleet as 1036TV, having become the 'turnover bus', being repeatedly turned on its side to train crews to recover any buses which overturned as a result of accidents. Also in this category, RT 19 had been selected for use as the chassis in the Chiswick training school in April 1954, being allocated the number 1020J later in the year. This was logical, since RT 19's chassis had been brought to post-war specification, thus being regarded as representative of the post-war RT fleet, then at its peak of 4,674 vehicles. The body, 18246, originally built for RT 1, also survived, being used for the mobile instruction unit 1019J, based at first on a former SRT chassis and then on that of a Cravens RT whose original body was damaged beyond economic repair. With this 1939 prototype RT body, the latter became 1037J.

2RT2 buses were involved in the service from Morden Station to Epsom Downs for the last time during the 1955 summer meeting. Pictured here operating the service on 25th May 1955 is RT 95, which would be withdrawn from passenger service at Chelverton Road garage at the end of that month. It was on loan to Putney Bridge garage for the occasion. *N Rayfield*

Withdrawn 2RT2s in store at New Cross garage in September 1955, all of which had spent their last days in service from one of the two Putney garages.
Denis Battams

A restricted blind layout was installed on RT 79 for the second time in its career upon the vehicle's transfer to the Country Area. This was to enable blinds from a departing 18STL20 to be used before sets of new blinds arrived for the bus and the other green 2RT2s similarly modified. RT 79 was to become the last 2RT2 training bus, being relieved of this duty on 13th February 1963.

With the exception of RT 62, the 2RT2s selected to operate in the Country Area came from Putney Bridge's last 29 withdrawn from service on 16th March 1955. RT 114 was transferred to Hertford on 23rd May. In this view it is still masked for a restricted intermediate display despite having full blinds.

Photographed displaying blinds for the Hertford garage based routes on which they often worked are RTs 114, 62 and 137. Although officially allocated to the garage to work only on the 327, the 'magnificent seven' green 2RT2s (and occasionally red RT 133) were regularly seen on routes which did not cross the weak bridge at Broxbourne, this structure being their raison d'étre at Hertford.

However, as it turned out, seven of the class were to re-enter passenger service. This was because of a weight limitation on Country Bus route 327, which was being operated from Hertford garage with some of the final batch of STL-type buses based on the immediate post-war standard AEC Regent II chassis with 'provincial'-style Weymann bodywork. The problem was a bridge at Broxbourne which had a load-carrying restriction not high enough to permit a standard post-war RT with a full load of passengers to pass over it. Work was to be done to strengthen the bridge, but was not due until 1957. The STL buses were lighter and still serviceable but had already found buyers following the decision to sell them off with the rest of the class.

It was therefore decided to employ the seven best 2RT2s, those selected being RTs 36, 62, 79, 93, 114, 128 and 137, all of which had been previously updated to the 3/2RT2/2 specification during the 1948-51 conversion programme, having been returned to service from overhauls that had been completed in October 1954. With the exception of RT 62, these survivors were drawn from the 29 members of the class taken out of service at Putney Bridge during mid-March, at which establishment they remained until transferred to Hertford garage (code HG) during the last two weeks of May. As a prelude to the arrival of the 'magnificent seven', RT 30 transferred from Putney Bridge to Hertford on 16th March, being relegated to learner duties in the process but remaining at the Country Area garage in this role long after its compatriots had departed.

The vehicles were repainted into Country Area green livery, thus becoming the first green examples of the 2RT2 type, apart from the one brief exception of RT 97 for a few months in 1946. Photographs taken of these 2RT2s in the first few months of operation on the Hertford-based routes show at least four equipped with restricted blind layouts, for the second time in their existence, in order to utilise the blinds from the 18STL20s which they replaced. The vehicles began work on 1st June 1955, thus maintaining continuity of the type in passenger service following Central Area withdrawal the previous day. One additional bus, RT 133, also ran from Hertford on two separate occasions: the first in August 1955 and the second from June to September 1956 and, although PSV licences were issued to cover these periods, the actual designation of the vehicle at the time was that of 'spare staff bus'. RT 133 was never repainted and thus became the last red 2RT2 to run in revenue earning service for London Transport.

In the event, the bridge reconstruction was delayed for a few months but was completed in time to allow the seven green 2RT2s to be withdrawn from London Transport passenger service for the last time on 31st August 1957 at the age of 17 years – a creditable age for buses which had a difficult start to their lives, operating in wartime London with its inevitable fall from London Transport's normal standards of good care of its rolling stock.

The First RTs

Left All green 2RT2s re-entering service on 1st June 1955 were examples of the 3/2RT2/2 conversion which had been applied to 99 of the type between 1948 and 1951. RT 128 was amongst the first vehicles to receive the modification, which occurred during its second overhaul at Chiswick between 16th July and 13th September 1948. *D.A. Jones*

Below left RT 36, the lowest numbered of the seven green 2RT2s, was the first to arrive at Hertford, this transfer taking place on 18th May. Each of the buses selected for further service had been the subject of a recent overhaul, all having been completed at Chiswick during October 1954.

Below centre RT 62 was the only 2RT2 to operate from all five garages which were associated with the type. Whether it actually operated on route 715 whilst allocated to Hertford is a matter for conjecture, though the destination display in this photograph was almost certainly set up just for the camera. However on 25th January 1956, RT 36 is recorded as having worked relief duty on this Green Line route.

Below The green livery applied to the seven 2RT2s made them suitable for working from other garages in the Country Area. Here RT 93 is seen on an unrecorded loan to Reigate.

CHAPTER EIGHT

Disposal and Afterlife

Apart from the handful which had been withdrawn because of fire (RT 85), accident (RT 22), or diversion to other uses (RT 19 as chassis to the Chiswick Training School, RT 97 to RTC 1 and RT 106 to turnover vehicle 1036TV), the 2RT2s might have remained in service for a year or two longer, had it not been for the unexpected reduction in London Transport's needs for buses as services began to be cut back because of the fall in demand. The decision to withdraw them from passenger service did not mean they had to be sold off, and as mentioned in the previous chapter no fewer than 127 were in stock after the end of 1955, mostly transferred to driver training or staff bus status, this figure also including the seven temporarily retained for passenger duty in the Country Area, all of which afterwards reverted to trainer or staff bus duties.

Already, many of the type had spent considerable time on training duties in between spells of passenger service, and in many ways they were almost ideal for the purpose. They looked like the standard London bus of the time and, more important, had driving controls and their response, and indeed the whole cab layout, almost identical to the post-war RT, yet were time-expired and hence their employment for such duties released more valuable modern buses. If no longer used for carrying the public, they could be licensed as 'private' and were no longer subject to public service vehicle requirements. London Transport still kept them mechanically sound but not always in smart order.

In addition, London Transport had a considerable need for staff buses, mainly to carry engineering staff to and from the various works, quite often on the basis of long-standing arrangements set up when the premises were opened and staff transferred from other parts of London. There had been a sharp increase in the need for such vehicles as the Aldenham bus overhaul works came into use. It was a considerable distance from urban London and almost impossible to reach by public transport. Many workers had been transferred from Chiswick to Aldenham as the latter expanded, so large numbers of staff had to be conveyed there from the area around Chiswick. For the first few years, many 2RT2 buses swapped roles from time to time, with spells of training and staff bus duties alternating.

Chiswick Works was responsible for dismantling RT 59 in June 1955 although the reason for this action remains unrecorded. It could be speculated, however, that the 1944 V1 incident, in which the vehicle was severely damaged, had been responsible for a number of latent body weaknesses that were not corrected either by Birmingham Corporation or in the three complete overhauls that followed.

Seventeen 2RT2s were sold in December 1955. It may well be that this was a case of London Transport dipping its toe into the water, as it were, to assess the potential market, bearing in mind that it was clear that it had more buses than it was going to need. It had already been decided that the 120 RT buses bodied by Cravens and dating from 1948-50 were to be sold off.

RT 15 was prematurely retired to the training fleet during March 1954 at New Cross, but just two months later was returned to passenger service from the same garage in which capacity it was to remain for a further twelve months. Demoted for a second time at the beginning of June 1955 the vehicle was used for driver instruction and operated from Alperton, Merton, Norwood, West Ham, Mortlake and Chalk Farm before being sold to F. Ridler in April 1963.
LT Museum

This July 1955 photograph shows RT 117 as adapted for its new role as a training vehicle with the glass behind the driver's seat removed. The full complement of post war moquette-covered seats stems from the fact that the bus ran right up to the last day of the type in passenger service at Chelverton Road just a few weeks earlier. *Alan B. Cross*

Below left Although officially allocated to Riverside garage in May 1955 (and later to Hounslow garage) RT 98 was a regular performer on the Chiswick skid pan. Photographed on countless occasions, RT 98 featured in an advertisement film for Wrigley's chewing gum and a 'Look at Life' interest feature for the cinema before being finally sold for scrapping in 1962.

Left RT 44 was among the first contingent of 2RT2s to be allocated to Chelverton Road and licensed for service from 2nd January 1940. Unlike the other vehicles which formed the original batch, RT 44 uniquely remained in service from the two Putney garages for the whole period of 2RT2 operation in the Central Area. During its later life the bus served as a staff bus and a training vehicle and is seen here in the latter role on an outing from Hendon garage where it was allocated from 1960-1962.

When the first bulk sale of seventeen 2RT2 buses was made to W. North Ltd, the Leeds dealers, in December 1955, they all saw further passenger service with independent operators. Smith's Luxury Coaches (Reading) Ltd purchased four and they were to be seen on that concern's contracts conveying workers to and from the Aldermaston and Harwell atomic energy establishments until the early sixties. Not only were they always smartly turned out in the distinctive blue and orange livery, for the managing director, Mr A.E. Smith, took great pride in the appearance of the fleet, but they also had the benefit of a well-equipped maintenance workshop. The former RT 3 is seen with the platform door fitted in 1959 – note the polished rear wheel disc – while RT 87 had non-opening windows installed at the front and its roofbox converted to a ventilator.
M.R.M. New

Although many major operators had caught up with their replacement programmes, and preliminary enquiries about the Cravens-bodied buses within the British Transport Commission empire had revealed little interest in even relatively modern ex-London vehicles, there was still a demand from various independent operators for serviceable buses. The modern-seeming RT was clearly an attractive proposition in terms of passenger appeal, though small concerns were apt to be nervous about what to them seemed in those days quite a complicated bus with unfamiliar air-pressure systems. Indeed, at that date, few of the larger company fleets had shown much interest in epicyclic gearboxes, and air-pressure brakes were only just beginning to become widely adopted.

Eleven of the batch of seventeen 2RT2 buses sold in December 1955 to W North Ltd of Leeds were examples of the earlier modification made to the air pressure system for which the code 1/2RT2/1 had been allocated, although it was not until September 1963 that the last of the type bearing this classification was sold. North, a dealer, was well-known as a buyer of time-expired vehicles from major fleets, including many earlier London disposals (RTC 1 had passed through its books the previous March). The new owners which this firm found for every vehicle in the batch, all independent operators, were not particularly well-known names for the most part, being either relatively small concerns with a service on which double-deckers were justified, or larger operators engaged in conveying workers to major contract sites.

In the latter category was Smith's Coaches, of Reading, which took four vehicles by May 1956, these being RT 3, 54, 87 and 108, but this concern was unlike most suppliers of transport of this kind in taking considerable pride in their turn-out and long-term maintenance, painting its vehicles in a smart blue and orange livery. All four ran until the early 'sixties, RT 54 and 108 receiving platform doors by about 1960. RT 54 was not withdrawn until April 1966 and, after a period in store, passed into private preservation in January of the following year. RT 108 was sold in December 1964 to L. F. Bowen Ltd, of Birmingham, passing to a subsidiary, Arnold's Coaches, of Twogates, Tamworth where it ran until withdrawn in January 1967, and is believed to have been the last 2RT2 to run in normal passenger service with an operator. RT 3 was withdrawn by September 1963 and sent to breakers the following May. RT 87 was out of use by November 1964 and was used for spares before being sold for scrap in May 1965.

A. Harris, using the fleetname Progressive Coaches, of Cambridge, purchased five vehicles in January 1956 – RT 32, 40, 76, 84, and 139, but seems to have acted as a dealer itself, to some degree. The following month RT 32 was sold on, to the London Brick Co Ltd, operators of a large fleet of AEC lorries. RT 40 ran for Harris until 1959, being sold to A. S. Philips (trading as B&P Services) of London NW2, operating on hire to a contractor. RT 139 was sold in March 1960 to a Cambridge contractor and in July 1961 to a local breakers. Of the others RT 84 is known to have received platform doors.

Above Concluding the quartet of 2RT2 buses sold to Smiths of Reading is RT 108 which, like its companions, was fitted with manually-operated rear doors during 1959 and 1960. The smart interior refurbishment, with heater, is also illustrated. This vehicle actually saw service with two other transport concerns following its withdrawal by Smiths and went on to become the last 2RT2 in normal stage service with an operator. *B.R.P. Moore*

Left After completing about four years service with Progressive Coaches of Cambridge, RT 139 found a second home, in March 1963, with Johnson and Bailey who were building contractors in Cambridge. The company sold the bus for scrap in July 1961. *G. Mead*

Beeline Roadways, of West Hartlepool, was another coach operator also running a number of double-deckers, and purchased RT 25, 125 and 149 in 1956. All were put up for sale in 1958, when RT 25 was sold to S. & H. M. Connell, 'Ubique Coaches', of London W9, for whom it ran until November 1963, losing its roofbox in the process and eventually being sold to breakers in February 1964. The same firm bought RT 125 after withdrawal in October 1959 but did not collect it, so it was scrapped by Beeline. RT 149 was sold to Rodgers, Station Garage, Redcar in May 1958, by whom it was sold for scrap in 1961.

Possibly the best-known operator of 2RT2 buses after their sale by London Transport was Red Rover, of Aylesbury, who acquired RT 31 and 61 in January 1956, painting them very smartly in its maroon livery in a style which gave them a look very close to that of RT buses still in service, save for the darker shade, and numbering them 8 and 7 respectively. RT 31 was withdrawn in September 1959 and sold to Kirkby, the dealer, of Anston, Sheffield, but RT 61 ran until November 1961, being broken up by Red Rover in June 1963.

Left Still carrying its distinctive front dome extending beyond the front windows, RT 25, now minus its front roof box, stands in the yard of Ubique Coaches in Portnall Road, Maida Vale. The vehicle returned to the capital in May 1958 from County Durham where it had been operated by Beeline Roadways of West Hartlepool since its acquisition from W. North of Leeds. Ubique used the vehicle for five years, mostly on contract work.
Colin Brown

Below Two 2RT2 buses that seemed particularly to catch the imagination of enthusiasts after sale were those purchased by Red Rover Omnibuses Ltd, of Aylesbury in January 1956. This may have been partly because they were used on a stage service in a district at the fringe of the London Transport area. Both lost their roof-mounted route number boxes but were smartly turned out, as can be seen in this view of RT 61, which ran with Red Rover for almost six years.

The remaining buses sold via North's went singly as follows: RT 2 was purchased by Yuille, of Larkhill, passing by 1958 to Ayrshire Bus Owners ('A1 Service'), of Ardrossan, the owner being H. H. Steele, of Stevenston, by whom it was withdrawn by July 1959. The A1 co-operative became a major user of post-war ex-London RT buses. RT 27 went in January 1956 to Anderson Bros. of Evenwood, Co. Durham, passing to Bird's of Stratford in May 1958. RT 140 was sold in August 1956 to C. J. Smith ('Bluebell Coaches') of March, Cambs. When replaced by a post-war RT, it was sold in July 1959 to Les Owen Tours, of Walsall, being withdrawn in 1961 and scrapped.

The seven buses temporarily retained for Country Area duty and painted green were added to the staff and trainer bus fleet when taken out of public service in 1957, and the overall situation remained virtually unchanged until 1959, when it was decided to withdraw many of the type. There was also a general pattern of change of status for those vehicles on staff bus duty which it was decided to retain, as they were reclassified as trainers from September 1959.

The next sale of a 2RT2 from LT's fleet occurred in August 1959 when RT 73 was sold to Mr J. Cochrane, effectively reducing the number of the class in stock to 126 from the figure of 127 which had existed since December 1955. The vehicle was purchased on behalf of a party of 36 Oxford undergraduates for a trip to Moscow and Leningrad, made during the same month

after those nominated to be drivers had received some brief tuition at the Chiswick Training School. An account of the trip by Viscount Walmer appeared in the April 1960 edition of Buses Illustrated in which he described the considerable difficulties encountered due to poor roads and low bridges. The outward journey was via Germany and Poland and the return via Finland, Sweden, Denmark, Germany and Ostend during which the vehicle gave no trouble despite travelling at what was reported as the 42mph governed speed wherever conditions permitted. On arrival back in Britain, the bus was sold in October to Pakamatic Ltd as a mobile showroom for the refrigerators which the company manufactured.

A total of 55 2RT2s passed from LT's ownership in 1960, representing the second largest clear-out of the class in any twelve-month period. The calm before the storm occurred in February when two members of the class were sold. The first, RT 74, had its body removed and scrapped by George Cohen at the yard then in use for dismantling the capital's trolleybuses at the rear of Colindale depot, its chassis then being presented to the Metropolitan Police College at Hendon. The chassis was returned to London Transport in September 1968 and was acquired by the London Bus Preservation Group soon afterwards. The second February sale involved RT 56 which was acquired, less all of its running units, by the Royal Earlswood Hospital in Redhill, Surrey for play therapy.

Above left RT 27 was a member of the initial batch of seventeen 2RT2s sold in December 1955 to W North of Leeds who found it a new owner in the shape of Anderson Brothers of Evenwood, near Bishop Auckland, who operated the vehicle until 1958. With no further buyers in the offing, the vehicle was sold to Birds Commercial Motors of Stratford-upon-Avon where it was scrapped in July of the same year.

Above RT 73 leaves Chiswick Works at the start of its trip to Moscow with 36 Oxford undergraduates in August 1959. *The Press Association*

Above Scenes such as this at Bird's Commercial Motors of Stratford-upon-Avon were repeated at a number of sites up and down the country as the withdrawal and wholesale scrapping of the 2RT2s got under way, the largest quantity going to this Warwickshire-based company.

Above right Ayrshire Bus Owners (A1 Service), an association of small independent operators, ran two 2RT2 vehicles at different times. RT 2 was acquired by owning member R.B. Steele in 1958 and was withdrawn about a year later whilst RT 58 remained with owning member Mrs R. Meney for approximately the same length of time, entering service in late 1961 and being withdrawn by August 1962. Meanwhile various other members had purchased post-war RT and RTL buses, as evident in this view – there were over 30 in use by A1 members in the early sixties. The style of painting, in blue, maroon and white, emphasised the differences in cab design between the earlier and later RT family variants.

The majority of the vehicles sold during 1960 were those acquired by another well-known dealer, Birds Commercial Motors of Stratford-upon-Avon to which establishment 51 of the type departed in a series of sales which commenced at the end of March and concluded in early December. Those sold in that first month were RTs 17, 105, 124 and 126. Following in April were RTs 26, 28, 57, 58, 103 and 143; in September (the main month) RTs 4, 6, 7, 8, 9, 12, 14, 16, 18, 20, 23, 24, 29, 35, 37, 38, 43, 46, 48, 49, 50, 51, 66, 67, 68, 69, 71, 91, 93, 99, 100, 101, 112, 117, 122, 135, 138, 145, 146, 148; and in December RT 64.

By that period, 20-year-old double-deckers had little hope of sale to operating concerns of any kind, and, in particular, what market there might have been had vanished as the result of the sale not only of the Cravens RTs but post-war RT and RTL type buses with standard bodies. Such buses had been snapped up by various independent and municipal fleets and inevitably put the 2RT2 in the shade with their relative youth and sturdier bodywork. Most of the 2RT2 buses sent to Bird's ended in the company's large scrapyard, though examples were offered for sale, and a few found new homes, but often more related to enthusiasm of one kind or another or specialised purposes rather than for normal use as buses.

Thus RT 8 went to Chicago for the British Week in that city in October 1961 and the following month went to the National Museum of Transport in St Louis, Missouri.

RT 17 went to Ferodo Ltd of Chapel-en-le-Frith, Derbyshire for brake testing duties until March 1965, after which it was sold in February 1966 to the 2RT2 Preservation Group for spares.

RT 57 went to Sweden in May 1960 to a plywood manufacturer as a mobile demonstration vehicle, and was reported with the MHF Motor Organisation, Sweden in 1965.

RT 58 was sold in August 1960 to A1 Service, Ardrossan, being withdrawn in August 1962. In October 1963, it returned to Bird's and was scrapped some years later.

RT 103 was acquired by St Michael's School, Ingoldisthorpe, Norfolk in September 1960 and in May 1963 was loaned to a holiday camp some four miles away at Heacham. Out of use by September 1964, the bus went to a local breakers in February 1966.

RT 112 was seen at the Le Mans 24-hour sports car race in June 1963, re-registered 999-EG-01, but its subsequent fate is unknown.

RTs 124 and 126 were purchased by B. Davenport, 'Blackheath Coaches', of Netherton, near Dudley in May 1960. RT 124 kept an appointment with an unknown scrapyard following the vehicle's withdrawal in June 1962 but RT 126, which had been cannibalised for spares in 1961, was later restored to full working order, re-entering service in 1964. This resurrection was brief and the vehicle presumably went the same way as RT 124 when disposed of in January 1965.

RT 138 was sold to the American Military International Insurance Association (AMIIA) and used for a prestige tour to Frankfurt, Germany in early 1961. During 1970 it was discovered in use as an office for a firm of car insurance brokers parked outside the USAAF base at Spangdahlen near Bitburg, where it was later sold to a painting contractor. Although no current information is available the vehicle was believed to be still in existence 25 years later.

Above left Acquired from London Transport by Birds Commercial Motors, RT 57 found a new owner immediately in the shape of Board Och Plywood Producer who exported the vehicle to Sundsvall, Sweden in 1960 as a mobile demonstration vehicle for their product 'Escaboard'.

Above Pictured at the Le Mans 24 hour race event held in June 1963 is RT 112 which was purchased from Birds Commercial Motors by a group of students and transported to France. It will be noted that the vehicle had been re-registered 999-EG-01. The front blind box has been removed and the body has been the subject of a number of small modifications.

Left Sold to the American Military International Insurance Association, RT 138 was used for a prestige tour to Frankfurt during 1961. It is seen in Greenford in February of that year before departing. It was later reported as being outside the United States Airforce Base at Spangdahlen in the West German Eifel Mountains and although this was confirmed, there has been no recent information regarding its fate. *B.R.P. Moore*

RT 143 (the initial 1/2RT2/1 conversion), with body removed and scrapped, was sold in January 1962 to the Limmer and Trinidad Lake Asphalt Co Ltd, of Fulham, London, users of a number of ex-passenger AEC vehicles for road resurfacing. The original engine, radiator and gearbox was removed and replaced by a 9.6-litre goods vehicle engine, radiator and four-speed crash gearbox to form the basis of a mobile street patching unit, numbered 437 in its southern fleet. In 1972 it passed to the Tarmac Construction Co, which had taken over the Limmer business.

The two remaining sales during 1960 involved RTs 131 and 136; RT 136 being acquired by F. Ridler of London W10 in June and passing immediately to another company for dismantling. In October RT 131 was purchased directly from LT and joined RT 56 at the Royal Earlswood Hospital in Redhill, the later arrival encountering the breaker's torch in Epsom in June 1966; RT 56 was scrapped on site in June 1969.

Only two 2RT2s were sold during 1961. RT 134 was acquired in January by Ridler, and probably went straight to another concern for dismantling as in the case of RT 138 some six months earlier. Ridler would later become the major purchaser of the 2RT2s yet to be declared available for disposal. The remaining 2RT2 sale in 1961 occurred in February when RT 129 was purchased by Botley's Park Hospital, Chertsey. Unfortunately the vehicle was burnt out around October 1963, its remains going to an unknown breaker shortly afterwards.

There was a general lull in 2RT2 withdrawals until August 1962 when the direct sale of RT 89 heralded the start of the final round of disposals; eight departing LT stock before the year was out. Then 1963 saw 59 vehicles sold off and 1964 the remaining two. In general the buses were sold to dealers for scrap, names that recurred being F. Ridler of London W10 and later of Chertsey Road, Twickenham; L.W. Vass of Ampthill, Bedfordshire; W.M. Gore of Wanstead E11; George Cohen Ltd, of London W6; Lammas Motors Ltd of Battersea SW17 and, in the case of RT 10, Barking Metal Trading Co. Not infrequently, there was then further movement to other dealers or scrapyards, Gore and Lammas passing their purchased vehicles on to Mountnessing Autospares in Essex. Ridler, on the other hand, dealt with a number of yards, most notably that of J.W. Hardwick in Ewell and another that was to be found at the rear of V. Collins' garage at Wrotham in Kent. In fact Hardwick's yard was the final site in the country where 2RT2s could be seen in any quantity, the last examples being dismantled there during 1974. Four of Ridler's acquisitions finished up in Norfolk with Richardson and Sons of Attleborough. It was here that the bodies from RT 55 and RT 114 were scrapped before their chassis passed to Culling and Son at Claxton, ostensibly for spares. RT 81, however, was cut down and became a flatbed lorry for Richardson's own use whilst RT 121 remained complete and served as a messroom for peapickers at a farm in Flaxton.

The First RTs

Even so, amid all this carnage, there were still instances of survival:-

RT 21 was sold by Lammas via Mountnessing to the Aztecs Beat Group in January 1964.

RT 42 was sold to C. S. Larkin, of Larkin Forge, Chesham, in May 1963.

RT 77 was sold by London Transport in December 1962 to the Portland Engineering & Grading Company of Crowland, Northamptonshire, but then on to J.W. Banks, of Witham-on-the-Hill, Northants and used for school contracts, being fitted with platform doors. Its condition gradually deteriorated until it became very dilapidated and ready for the breakers in 1996.

RT 89 was sold direct in August 1962 to the Hamsters Mobile Theatre Group, of East Ham, E6. In April 1965, it was damaged by vandals but sent for repair to Audawn Coaches, Corringham. The Hamsters Group sold it in May 1967, presumably for scrap.

Facing page top Except for its registration number, there was little to identify RT 143 when Limmer and Trinidad Lake Asphalt Co Ltd had finished its conversion following its purchase in 1962, for it took on an appearance much like an AEC goods vehicle of the late pre-war or early post-war era. Indeed this was more than skin deep, for there was a standard goods-type 9.6-litre engine and crash gearbox behind the taller radiator. It seems that the lower built passenger frame was chosen to accommodate the tar boiler and other equipment required for the street patching unit, as Limmer and Trinidad Lake had a number of older AEC Regent chassis which had received similar bodywork.

Facing page centre Neglected and slowly disintegrating, RT 56 stands minus its running units and awaiting an appointment with the scrapman at the Royal Earlswood Mental Hospital at Redhill in Surrey in June 1969. The bus had been purchased directly from London Transport in 1960 and was used for play therapy by the patients. The hospital also owned RT 131. *N. Rayfield*

Facing page bottom The conversion of RT 81 for use as a transporter was carried out by Richardson and Sons of Attleborough in Norfolk who acquired the complete vehicle with three other members of the class during the summer of 1963 from F. Ridler (Dealer) of Twickenham, Surrey. *B.R.P. Moore*

Top right Following withdrawal, RT 21 was acquired by Lammas Motors of Wandsworth who subsequently sold it to Mountnessing Auto Spares in Essex. This company acquired and scrapped a number of 2RT2 vehicles but this example was sold on to the Aztecs Beat Group who painted it yellow and named it 'Brigitte'. Last seen in August 1965, its final fate remains unknown. *G. Mead*

Right RT 89 was sold directly by London Transport to the Hamsters Mobile Theatre Group in August 1962. Unfortunately, the vehicle was later the subject of an attack by vandals in April 1965 and was consequently sent to Audawn Coaches of Corringham for repair. Photographed on 28th August 1965 after the repairs had been undertaken, it will be noted that the bus has been fitted with a windscreen and side cab window from a post-war RT. *G.R. Mills*

Probably the most maltreated 2RT2 was former RT 106, which in November 1955 became a turnover vehicle for LT staff to practise righting. Renumbered 1036TV, it could often be seen in later years in the grounds of Stonebridge garage, where it was photographed. Another bus used for the same purpose was RT 119, when allocated to New Cross, between November 1962 and October 1963.

Facing page RT 1's body belies its age in this view of 1037J taken on 21st July 1967, whilst the bus was based at Wood Green garage. Unfortunately the vehicle's condition was allowed to deteriorate in later years and it left London Transport in a very sorry state when sold for preservation at the end of 1978. Although the chassis, from RT 1420, was still officially registered JXC183, 1037J's movements between garages were always accomplished using trade plates. The blinds which came from Putney Bridge remained fitted for some time and these, together with RT 19's rear registration, gave some indication of the body's previous history.

RT 111 was another example of a direct sale which took place in April 1963 when the bus was acquired by the Civil Defence Officer in Croydon. The bus then added some realism to the Civil Defence training ground in Albert Road, Croydon where it was used in many exercises until it was cut up on site in January 1968.

The last sale of 1963 occurred on 6th December when RT 119 was acquired by Colbro Ltd of Leeds who passed the vehicle on to C. Hoyle of Wombwell, Yorkshire, for breaking; an area that would become synonymous with the dismantling of many post-war RT variants.

The remaining two 2RT2s, RTs 88 and 118, held in stock at the beginning of 1964 were finally sold at the end of March to George Cohen Ltd, both suffering a lingering demise at the company's Canning Town premises.

Yet still one complete vehicle, one body and one chassis from the first RTs existed in LT's fleet. The complete bus, formerly RT 106, carried the number 1036TV in the miscellaneous stock fleet – its raison d'etre since November 1955 being to demonstrate the art of lifting overturned buses, TV signifying

Turnover Vehicle. Finally, in 1970, the role of this much-maltreated vehicle was usurped by a post-war RT and it was subsequently sold for spares to the London Bus Preservation Group in March 1971.

The chassis and body remaining in stock following the sale of RTs 88 and 118 had been united from 1945 to 1954 as RT 19, their separation resulting in the chassis transferring to the Chiswick Training School to assume an instructional role, becoming numbered 1020J in the process. Its sale to L.W. Vass for scrap in October 1965 seems a pity for, as the chassis passed into oblivion, the chance of reuniting it with the still-extant prototype body, was lost for ever.

And so it was that the first became the last, for the body which had been built for the prototype in 1939 still soldiered on, paired with its fourth chassis, in the guise of Mobile Instruction Unit 1037J. It was declared surplus to requirements in 1978 after which this unique hybrid was acquired by Prince Marshall's Obsolete Fleet and restored; the body regaining its original fleet number, RT 1, and the chassis being re-registered EYK396, so as to simulate the prototype.

CHAPTER NINE

Preservation

In October 1962 four bus enthusiasts gathered to discuss the possible purchase of a member of the 2RT2 class from London Transport. At that time there were 61 of the type remaining in LT's stock but their days were numbered and 1963 would see the disposal of all but two, this action representing the largest withdrawal of the class in a single year.

The 2RT2 Preservation Group was formally established on 2nd November 1962 (to coincide with the first letter written to London Transport enquiring about the possibility of purchasing a bus), the original membership comprising Brian Moore, Sidney Hagarty, Francis West and Derek Parsons, the only member with a road transport background, being employed as a bus driver for London Transport at Holloway (J) garage. Despite their strong commitment to acquiring a 2RT2 for preservation, only Brian Moore had been previously involved with a similar project. This was the aborted attempt in 1954 by another group to purchase STL 2674 from London Transport; but these embryonic preservationists, which included such worthies as Alan Cross and Prince Marshall among their number, had failed to raise the asking price.

The response from London Transport to the 2RT2 Group's letter was curt and to the point stating that, as soon as some suitable vehicles were available the Group would be given an opportunity for tendering. But here were uncharted waters; the Group knew little about making offers for London Transport's redundant road vehicles and a further letter was sent to Chiswick asking whether an exception could be made. The reply from the Supplies Officer was quite informative and stated that the price of a vehicle fitted with slave tyres would be between £125 and £200 according to condition. The letter also contained the following statement: 'Vehicles are sold as 'runners' but without any guarantee whatsoever, and I would remind you that certain spares for the RT2 type bus would be almost impossible to obtain in the event of a major breakdown.'

In conclusion the letter stated that the Group would be given an opportunity of making an offer for a vehicle, the Executive being unable to deal with individual requests. The only means of ensuring rapid disposal was for batches of vehicles to be inspected and tendered for simultaneously.

And, true to its word, London Transport provided a tender form for six of its vehicles (RT 83, 98, 110, 116, 118 and 128) which arrived for consideration by the Group on 29th December 1962. It was then the Group members made an important and timely decision for they realised that, as a considerable number of the class still remained in stock, they would not only be able to inspect those vehicles on the tender form, they would also have the opportunity to evaluate any other 2RT2s which had yet to be officially withdrawn.

On Sunday 6th January 1963, the four founder members ventured forth to such diverse places as Battersea, Upton Park, West Ham, Hammersmith, Stockwell and Shepherd's Bush to give the aforementioned vehicles the 'once over' with the additional incentive of being able to examine further examples of the class at close quarters.

On his return, Francis West typed up a list of the vehicles inspected and their condition. Examined today, this provides an interesting analysis of the final state of some of the 2RT2s in stock at that time.

Vehicles at Upton Park garage (U):

RT 116 FXT291 06616864 b.384 (Uu) (Listed on tender form);
Not considered, as very badly damaged platform due to collision. Bodywork fair.

RT 127 FXT302 06616875 b.394 (Ut) (Not listed on tender form: still licensed as Upton Park trainer);
Not suitable as fitted with non-standard, quarter-top windows. In reasonable condition, bodywork good.

RT 133 FXT302 06616881 b.375 (Ut) (Not listed on tender form: still licensed as Upton Park trainer);
Not considered as very bad damage to offside, probably collision damage. Bodywork poor.

Vehicles at West Ham garage (WH):
RT 36 FXT211 06616784 b.334 (WHt) (Not listed on tender form: still licensed as West Ham trainer);
Green livery. Reasonably good, bodywork fair. Destination box at rear end minus certain fittings. Sound.

RT 63 FXT238 06616811 b.315 (WHt) (Not listed on tender form: still licensed as West Ham trainer);
Bodywork poor, internal condition doubtful. Good pre-war seating.

RT 92 FXT267 06616840 b.385 (WHt) (Not listed on tender form: still licensed as West Ham trainer);
Very bad body sag, excessive rusting throughout. Not worth considering.

RT 118 FXT293 06616866 b.399 (WHu) (Listed on tender form);
Not worth considering. Extremely bad bodywork, eaten away with rust. Not too sound mechanically, believed to have been cannibalised for spares. No good.

Vehicles at Stockwell garage (SW):
RT 70 FXT245 06616818 b.397 (SWt) (Not listed on tender form; still licensed as Stockwell trainer);
Not much good. Poor bodywork with traces of rust. Very bad dent on roof panels.

RT 83 FXT258 06616831 b.377 (SWu) (Listed on tender form);
Traces of rust on bodywork. Very weak platform. Poor condition.

Vehicles at Battersea garage (B):
RT 110 FXT285 06616858 b.341 (Bu) (Listed on tender form);
Extremely bad bodywork. Not original roof route number box as fitted with a rebuilt one after accident damage. Bent radiator. Several cavities in cabside. Not worth considering.

RT 128 FXT303 06616876 b.393 (Bu) (Listed on tender form);
Best condition vehicle noted to date. Green livery, Platform rather weak and rear destination boxes need attention. One broken seat frame on lower deck. Chassis in very good condition.

Vehicle at Riverside garage (R):
RT 111 FXT286 06616859 b.410 (Rt) (Not listed on tender form; still licensed as Riverside trainer);
Poor bodywork with excessive rust. No clips to platform fareboard position. Offside grabrail missing. No mirror top deck (over staircase). One cracked window (top deck). No fittings to via destination box at front. Offside route stencil. Good nearside wing. Offside rear wing fairing missing.

Vehicles at Shepherd's Bush garage (S):
RT 98 FXT273 06616846 b.349 (Su) (Listed on tender form).
Extremely bad condition. Bodywork poor; great expanse of rust. Too far gone to be considered.

RT 113 FXT288 06616861 b.347 (Vt) (Not listed on tender form; still licensed as Turnham Green trainer);
Reasonably good. Top deck interior excellent. Slight dent on offside quarter panels. No offside route stencil. Part of mud valance missing. Broken bonnet support lever. No information board brackets. Poor nearside wing. Very slight body sag. Rusty window surrounds to two nearside pans, lower deck. Very good seating. Commendable.

Of the six vehicles listed on the tender form, only RT 128 was anywhere near the Group's requirements and a letter was sent to London Transport stating that no offer would be made on this occasion but intimating that a more suitable vehicle may yet emerge.

A gap of two months passed before the next tender form arrived which detailed twelve 2RT2 buses for disposal. This time all were situated at Fulwell garage, thus facilitating inspection. On 17th March 1963, all members visited the former trolleybus depot, and Francis's notes, as detailed as ever, record their conclusions:

Vehicles at Fulwell garage (FW)
RT 5 FXT180 06616753 b.283 (FWu) (Listed on tender form);
Poor condition. Very bad nearside wing. 'T' bracket to platform bodywork. Damaged rear roof route number box. Offside stencil.

RT 15 FXT190 06616763 b.280 (FWu) (Listed on tender form);
Not considered as rear roof route number box removed. No offside stencil. Poor condition. External condition reasonable.

RT 39 FXT214 06616787 b.284 (FWu) (Listed on tender form);
Reasonable. No offside stencil. Pre-war front wheel hubs. Seating useless. Rear offside damage in collision. Patch on nearside. Nearside garage and running number code brackets missing.

RT 41 FXT216 06616789 b.309 (FWu) (Listed on tender form);
Poor. Offside stencil. No fareboard brackets. No handle to rear destination boxes. Very bad wings. Very poor roof and front cab window.

RT 45 FXT220 06616793 b.309 (FWu) (Listed on tender form);
Useless. No code brackets, no information board brackets. Hole in nearside of bodywork. No platform blind handles, no tread on nearside wing. Interior handrail missing below front lower deck window (offside). Fareboard brackets missing. Various parts missing from destination boxes. No offside stencil. Electric horn fitted. Bad knock on engine.

RT 52 FXT227 06616800 b.281 (FWu) (Listed on tender form);
Poor. Grab handle on cabside missing. Small hole in cabside. Nearside route stencil missing. No information board brackets. Poor wings, special hooter fitted. Front handrail missing on top deck. No nearside driving mirror. Broken seat frame on top deck (second offside from staircase). Offside stencil plate.

RT 55 FXT230 06616803 b.370 (FWu) (Listed on tender form);
No offside stencil plate. Fittings missing on front ultimate destination box. Cracked radiator. Very good nearside wing. No handle to platform destination box. Bodywork poor with traces of rust. No front handrails, lower deck. No platform fareboard brackets. 'T' bracket to bodywork at rear.

RT 70 FXT245 06616818 b.397 (FWu) (Listed on tender form);
Already inspected at Stockwell garage on 6th January 1963.

RT 81 FXT256 06616829 b.365 (FWu) (Listed on tender form);
Reasonable. No information board brackets. Glass still fitted behind driver. Very poor condition of rubber surrounds to windows. No keyhole to front roof route number box. Very poor roof condition. Pre-war style front hubs. Offside stencil. Very good batteries; engine note good.

RT 111 FXT 286 06616859 b.410 (FWu) (Not listed on tender form);
Already inspected by group at Riverside garage on 6th January 1963. Additional variation to condition being damaged front foglight.

RT 113 FXT288 06616861 b.347 (FWu) (Listed on tender form);
Already inspected by group at Shepherd's Bush garage on 6th January 1963. No change on condition of vehicle: still the best RT2 inspected to date.

RT 114 FXT289 06616862 b.413 (FWu) (Listed on tender form);
Green livery. Patch on nearside of top deck bodywork. No information board brackets. Pre-war front hubs. Dented radiator. Cracked glass in front roof route number box. No garage/running number code brackets. Glass missing from platform window. Platform fareboard brackets missing. Handle missing to rear destination box. One handrail missing from lower deck. Poor seating. Cracked windscreen. Bad gash in offside panels. Traces of rainwater in front destination boxes. Post-war wings. Not worth considering.

RT 120 FXT295 06616868 b.420 (FWu) (Not listed on tender form);
Poor. No offside stencil. Post-war hubs. Bad dent on radiator. No information board brackets. Poor nearside wing. Filler cap rusted. No fittings to front destination boxes and traces of rainwater leakage. Traces of bodywork sag.

RT 121 FXT296 06616869 b.414 (FWu) (Not listed on tender form);
Not considered as fitted with quarter drop windows. Pre-war style front hubs. Poor seating. One offside top deck window missing. Cracked fuelling point. No offside stencil. Fareboard brackets missing.

RT 130 FXT305 06616878 b.414 (FWu) (Listed on tender form);
Useless. Offside driving mirror damaged. Non-standard headlamp on nearside. Post-war filler cap. No nearside destination blind handle. Nearside lower deck front window winding gear useless. Bad beading on emergency exit. Rear destination boxes poor. Very doubtful condition of lower deck floor. Centre bulkhead pillar weak. No offside fairing.

RT 150 FXT325 06616898 b.428 (FWu) (Not listed on tender form);
Not worth considering. Bad dent on nearside of roof. Patch on cabside. Offside stencil. Offside corner of bodywork rotten. Hole in canopy. No handrail behind driver. Traces of rust on platform. No mirrors fitted over staircase. No handles to rear destinaton boxes. Poor condition window pans.

By now the group members had examined 27 vehicles which represented a little under half of the 2RT2s which were still in stock. There was no doubt that RT 113 was the most suitable vehicle available and now it remained for the all-important decision to be made regarding the actual amount to be offered to London Transport for the bus. In the end, each of the four members suggested a price for the vehicle and an average was produced which amounted to the then princely sum of £105.00. The tender form was duly completed and sent to the Supplies Officer at Chiswick on 20th March 1963. Now all the Members could do was to cross their fingers and wait.

On 29th March, a letter accepting the Group's tender of £105.00 for RT 113 was received from the Supplies Officer of the newly-created London Transport Board.

In the weeks that followed, correspondence flowed between the Group members and the various companies which had been associated with the development of the 2RT2 class in an attempt to obtain servicing documents. However a more pressing problem, now in the minds of all members, was the locating of a suitable parking space for the vehicle once it was in their possession. Many sites were investigated but most were unavailable or too expensive. At the eleventh hour an agreement was reached with Clayton Garage in Alperton where the bus could be parked in the open air at the rear of the premises for fifteen shillings per week. Now all that mattered was for the group to collect its prize.

It may have been by accident or design that RT 113 passed into the ownership of the group on 1st May 1963, which coincided exactly with the same date twenty-three years previously when the bus was first licensed at Putney Bridge garage.

From the correspondence on file it appears that, at the time of acquisition of the vehicle, the Group's intention was to carry out a restoration programme that would be as brief as possible and letters to paint manufacturers and advertising poster publishers would tend to confirm this commitment. Scrapyards such as those of George Cohen at Canning Town and Kettering, which were engaged in dismantling other members of the class, yielded a wealth of spares whilst nearside wings were exchanged with RT 111, which was then to be found at the Croydon Civil Defence Training ground.

The Date: 1st May 1963.
The Place: Fulwell Garage Yard.
The Occasion: the Collection of RT 113 by members of the 2RT2 Preservation Group.
Derek Parsons

All the while the location of the bus, at Alperton, was a closely guarded secret and, fortunately, Francis West, a PSV Circle sub-editor, as able to censor any reports that came in to him for possible inclusion in the Circle's Metropolitan News Sheet. But when two reports appeared in Buses Illustrated, the Group decided to respond by writing a letter to confirm the vehicle's acquisition.

In true Noah's Ark fashion, readers of Buses Illustrated would have noted two letters in the April 1964 edition both announcing the acquisition, for eventual preservation, of two members of the 2RT2 class. The first letter was from the Group in answer to the query raised in an earlier issue, the second from Mr E.H. Brakell who had recently purchased RT 44 from J. Hardwick, a dealer at West Ewell in Surrey. Mr Brakell's original intention was to use RT 44 as a store in connection with his building business but, in the event, the vehicle did not suffer this fate and went on to become a regular participant in many rallies soon after its acquisition.

By the time the letters appeared, the Group had already made contact with Mr Brakell and eventually arrangements were made for the two vehicles to meet. This occurred on 9th August 1964 when RT 44 was driven to Harringay Service Station where RT 113 had been based since the previous December, following the receipt of a notice to quit Clayton Garage at Alperton. RT 44 was to remain in the safe keeping of Mr Brakell until 1995 when the bus was sold for continued preservation to Mr R. Wood of Marchwood, Hants.

In the photographs taken at the meeting of the two buses, RT 113 looks much the worse for wear with the nearside of its upper deck covered by a tarpaulin. This was due to the fact that work was in hand with the reframing of the body as investigations had confirmed that the slight body sag noted at time of the Group's inspection of the vehicle was due to extensive rotting of the seasoned ash framework. Sidney Hagarty's carpentry skills were therefore being put to the test in reproducing a new framework, but this time in teak, the curve in the upright pillars being achieved by placing them over a bucket of hot water on a gas ring and letting steam and human body weight do the rest. Had this momentum in restoration work been maintained, it is quite likely that RT 113 would have been finished much earlier than it was. However, the Group's goal was to get the bus into covered accommodation and every effort was made in that direction. There was a minor flurry of excitement during late 1965 when it was learned that the Eastern National garage at Silver End in Essex was to be sold and negotiations for its purchase in connection with Passenger Vehicle Sales Ltd reached an advanced stage before the premises were withdrawn from sale.

Below left RT 113 at Alperton on 9th May 1964, shortly after the preservation project was made public knowledge.
Derek Parsons

Below RT 44 and RT 113 together briefly on 9th August 1964 at Harringay. RT 113, looking somewhat forlorn, is still in the throes of being re-framed. *Brian Moore*

As recorded at the beginning of this chapter, RT 113 was the first member of the RT family to be purchased for preservation in the UK, but the honour of preserving the first member of the class in the World falls, rather unexpectedly, to the National Museum of Transport in St Louis in the United States of America where RT 8 has resided since November 1961. This disposal would have gone without notice had a gentleman by the name of Alan Pommer not written to the Group with this information. But this was merely an aside in Mr Pommer's letter, his main reason for communicating was to announce his intention to purchase RT 82 from London Transport and ship the vehicle back to Boston. Following a lengthy exchange of correspondence, arrangements were made for the Group to collect RT 82 on Mr Pommer's behalf and drive it around the capital with a special emphasis on photographing the bus at Hyde Park Corner – Mr Pommer's address being Hyde Park, Boston. RT 82's last trip over part of its old routes occurred on 15th March 1964 after which it was taken to Dagenham Dock where it was loaded on board a ship bound for the New World.

The Group maintained contact with Alan Pommer and from time to time photographs and magazine cuttings arrived which showed RT 82 undertaking various duties on the other side of the Atlantic – on one occasion it was painted into CIE livery for a local Irish week. On 10th September 1966 the Group received an International Telegram requesting urgent assistance as the roof of RT 82 had come off second best in a confrontation with a low railroad bridge. Apart from sending out plans of the body to Mr Pommer and listing the remaining members of the class, there was little the group could do to help. The following year, Mr Pommer visited Britain to inspect RLH 21 before arranging for it to be shipped back to the USA (a case of 'don't raise the bridge, lower the bus'). During his visit he became a member of the London Bus Preservation Group. When the chassis of RT 74 was declared surplus to requirements at the Police College at Hendon in January 1969, it was acquired by the LBPG and sold to Mr Pommer, making its trans-Atlantic voyage the following September.

The constitution of the 2RT2 Preservation Group allows for there to be a maximum six 'owning members' of the vehicle at any one time. Every person achieving such status must pay the equivalent of a quarter share (£26.25) of the original purchase price of the bus into group funds. It was on these conditions that Maurice Bateman and Tony Beard joined the group in July 1965, the latter's initiation being the job of stripping off approximately ten layers of brown paint from the staircase.

All members of the Group belonged to the Historic Commercial Vehicle Club, and in November 1965 this organisation's newsletter carried a note that RT 17 was available for sale from Ferodo Ltd at Chapel-en-le-Frith in Cheshire. Sensing the vehicle would provide a good source of spares the Group made a successful bid of £35.00. The bus, which until two years earlier had been used to test brake linings, was brought to London in February 1966 and parked at the back of the Ilford branch of W.H. Smith where Tony Beard worked with Maurice Bateman. Maurice somehow talked the manager into allowing the vehicle to be stored in the staff car park, which at the time was under-used.

By now two of the Group members had jobs in the bus industry. Sidney Hagarty had become a bus driver working from Wood Green garage whilst Derek Parsons had become an employee of Redbridge and District, a subsidiary of Super Coaches of Upminster, which operated a bus route (No.19) between Chadwell Heath and Woodford Bridge. Another concern linked with Super Coaches was Passenger Vehicle Sales (London) Ltd and it was this organisation which helped secure a covered parking space for RT 113 in Eastern National's Canvey Island garage (now the Castle Point Transport Museum) – the bus taking up residence on Friday 29th April 1966.

Progress on the restoration of RT 113 then became painfully slow. In 1969 RT 17 was towed to Canvey Island, to be dismantled for spares.

RT 82 was acquired by transport-enthusiast Alan Pommer of Boston, Mass, USA in September 1963. At the request of the owner, the vehicle was taken on a filming trip around the sights of London by members of the 2RT2 Preservation Group just prior to its departure for America on 22nd March 1964. In this photograph the vehicle is seen, painted in CIE livery for a local Irish Week, standing outside Mr Pommer's home in Boston.
Alan Pommer

The Group then suffered two blows in consecutive years. In April 1973, Francis West died as a result of a brain tumour and so sadly never saw RT 113 in its completed condition. And in 1974 a notice to quit was received from Eastern National, thus instigating another search for suitable covered accommodation. Fortunately, after an intensive time spent looking at railway goods sheds and aircraft hangars, Tony Beard discovered a vacant garage about half a mile from his home in Chadwell Heath which had recently been vacated by a removal company and, as the site was due for redevelopment, was available at a peppercorn rent. It was here that RT 113 and T 504 were successfully garaged whilst the chassis of RT 17 was parked outside. There then followed a few months of frenzied activity during which two enormous doors were made for the building, and power and lighting were installed. Sadly RT 113 did receive the attention of local vandals before the doors were finished, but damage was relatively slight.

The Group then applied to the local council in order to get relief on the rates for the premises but its finance subcommittee considered that the Group was a commercial enterprise and insisted on charging the full business rate. It took the persuasive powers of a local councillor and a Member of Parliament to get the decision revoked. However, three years had, by then, elapsed and there was still an outstanding amount. Sadly, the Group opted to sell the family silver and T 504 was sold to the LBPG for approximately the amount of the outstanding demand. The space that the vehicle had occupied in the garage could now be sub-let to another bus owner in order to share the occupation costs to keep the establishment viable. Contact was therefore made with the owner of RT 141, Mr David Gomm, who leapt at the chance of sharing the premises. Mr Gomm had rescued the 2RT2, minus its engine, from J. Hardwick at Ewell and soon after purchased RT 54.

RT 54 had certainly had a varied history. Following its purchase by Smiths of Reading, in June 1956, the bus saw a further ten years of service with this Berkshire operator, who later fitted platform doors and a heater. The vehicle was then purchased by Alan Farrow for preservation and was later owned by Mr Malcolm Bowyer and Mr Brakell before passing to Mr Gomm. In more recent years the vehicle has changed hands on more than one occasion and in 1996 was up for sale with a £3,500 price tag – a plan to operate it around the City of York having failed to bear fruit.

Although RT 141's platform had been professionally rebuilt, its framework remained in an advanced state of disintegration. An engine was eventually obtained for the bus, being that from RT 44 whose own unit had been replaced by that from RT 106 (1036TV). This engine became available following the acquisition of the turnover vehicle by the London Bus Preservation Group in March 1971 in whose ownership it was later dismantled. In the end RT 141 and its 'engine' became the property of the Woolwich Transport Museum, whereupon it too was dismantled for spares.

It is unlikely that any preserved RT has had more owners than RT 54 which was initially purchased by Alan Farrow from Smiths of Reading in 1967. The first serious attempt to return the vehicle to London condition was made during Malcolm Bowyer's ownership during which the platform doors were removed and immediate post-war livery applied, complete with brown painted roof. The roof colour was later changed to red by a new owner and repainted brown again by another, when the bus received its wartime livery. The failure of some of the wooden framework has resulted in an obvious body sag and as such will require some serious attention in the near future. *Alan Moore*

RT 44, in the ownership of Ted Brakell, on the sea front at Brighton at the 1970 HCVC Rally in May of that year. *N Rayfield*

With RT 113 now garaged locally there was no excuse for a lack of working parties and all the stops were pulled out in an attempt to get the project completed. A significant link was made with London Transport whose senior craft apprentices at Aldenham undertook much of the interior finishing work. And then, in October 1978, just as the Group members were at last seeing a return for their labours, an incentive to finish the job arrived in the form of a letter from the LBPG which asked whether the bus would be available to appear at the '150 Years of London's Buses' event to be held in Hyde Park the following summer.

Another vehicle in which the 2RT2 Preservation Group became interested was 1037J, carrying the body of RT 1, and from 1966 onwards letters were regularly written to London Transport requesting information about its disposal. When the bus was finally released in 1978, the group made an offer for this historic hybrid but, in this instance, was outbid by Prince Marshall.

Prince's main interest was in pre-war vehicles and he had little enthusiasm for the RT, which had eliminated the varied collection of vehicles in use in London and, indeed, because so many of them were already privately preserved. He was however periodically lobbied by Tim Nicholson, Principal Engineer for the restoration of ST 922 and D 142, that the quantum leap represented by the RT deserved one of the type in his own collection.

The body of RT 1 had remained in use as an engineering training unit mounted on the chassis of RT 1420 – itself the victim of a low bridge accident in 1954. Although so very important, this vehicle – known by its service vehicle fleet number 1037J – was widely ignored by preservationists and enthusiasts. It languished at the back of West Ham garage, its historical significance a matter of apparent indifference.

Prince Marshall quietly secured it before announcing the fact to an incredulous Tim Nicholson. In the negotiations Prince had promised a fully working RT 1 for the celebrations surrounding the final run of the last RT – only some six months hence. Thus 1037J made a rare outing in the daylight under tow to LPC Coachworks at Hounslow before commencing another period indoors. The journey was not without incident – as Tim's AEC Matador made an abrupt stop at a set of traffic lights, Prince Marshall was unprepared, and the unbraked RT on the end of the tow bar skidded sideways, just missing the Matador and ending up alongside it still joined by the now-bent tow bar.

Once lodged at LPC Coachworks restoration proceeded in earnest. Its mechanical overhaul turned out to be promising. RT 1420 had only recently been overhauled before its accident and the mechanical units were all in good condition with low mileage – only the effects of long-term disuse were apparent. Many sub-assemblies were in excellent condition, as 1037J had been an engineering training unit.

Its fading paint was removed to bare metal, and panels were replaced as necessary. The remnants of the offside signalling window for the conductor (a glass panel halfway up the stairs on RT 1) were discovered and restored. The bus did not have a full set of seats but Prince had already ordered a set of RT2 seat frames for it and these were provided. RT 1's original moquette was faithfully reproduced and upholstered onto the right size cushions. However, these items were ordered at the early stages of the project before RT 1's lack of compliance with 2RT2 standards became evident. As a result RT 1's seat frames were fixed into position with a selection of spacers, packers and wooden blocks.

Laurence Sheffer – Prince's advertising salesman who had found Johnnie Walker and others to sponsor his vehicle restorations – was despatched to the original advertisers on RT 1 for some support and indeed consent. Of these, Alka-Seltzer, Brimay and Bisto all agreed and consequently their posters were faithfully restored to the vehicle by signwriting. Maples, however, no longer made carpets and so Maples *Furniture* was the legend on their poster, but painted in the style of the original.

Prince elected to paint RT 1 in the original livery of red, white and silver roof with the polished half-round mouldings which had been painted over in 1939 and had stayed that way ever since. LPC Coachworks Director Eric Chambers personally removed all the mouldings, stripped the paint from them and polished them. The flared mudguards were made of fibreglass from moulds taken from STL 2093. For practical purposes we 'forgot' to change the nearside mirror for the square one originally fitted on the grounds that greater vigilance was needed in modern traffic conditions and to damage the bus would be unforgivable. The post-war welded wheels did not look right on RT 1 and the pre-war riveted type were fitted, thus enhancing its appearance.

Making RT 1 ready for modern operation required some ingenious solutions. Tim reconfigured the lighting system so that the front lights also acted as trafficators and flashed when required. The switch for this was hidden by the driver's left side and was not visible. At the rear it had to have another light fitted at the bottom of the rear platform rail (in a design not dissimilar to RM 1 and RM 2) and together with the original one nearer the right formed a pair of tail lights with trafficator facility. It also gained a bulb horn in the position where RT 1420's old trafficator turret had been fitted.

Mechanical restoration proceeded but actual testing was severely limited by work on the topside – much to Tim Nicholson's protests. As the date for the Class V annual test drew near (in itself 24 hours before RT 1's debut at Barking) he became increasingly concerned at the impracticality of any road testing. Even engine testing brought howls of protests from signwriters, painters and others! Eventually it passed its test, but alarmingly would slip when in top gear and labouring – something more evident on RT 1 with its Country Area differential from RT 1420. (Some years later this condition became unacceptable and the top gear plate was replaced. As a result of this process it was discovered that RT 1420 had experimental steel and fibre top gear plates as opposed to the normal steel and bronze.) Late on the night of 6th April 1979 the restored RT 1 was more or less ready for the road – still tacky from paint and filled with the aroma of wood, glue, rexine and leathercloth.

The Saturday morning dawned and as enthusiasts were enjoying the last rites on various RTs in Barking some distinguished guests made their way to Victoria garage to be carried aboard RT 1 to the final cavalcade. Calamity struck when Tim Nicholson started the engine and discovered it was stuck in top gear – a condition making it impossible to build up any air or move the vehicle. Experts will know that this condition normally required a look into the box and often a gearbox change. However on this occasion – with the clock ticking – Tim took the unusual step of jacking up the back of the bus so the nearside wheel could turn. With one wheel free to turn, shafts turned and air pressure began to build. When sufficient air pressure was available, Tim took the dramatic and risky step of selecting and engaging a lower gear, and this miraculously dislodged top.

Back on all four wheels there was an immediate departure for Victoria with no long-term effects from the problem that morning save the annoying top gear slip when labouring. However as she reached Hammersmith she gave for the first time some indications of fuel starvation. Preservationists will know that long-term idleness does more harm to a vehicle than frequent usage and no amount of cleaning overcomes the effects of long-settled matter being disturbed. Fuel starving RTs can be run gently for a couple of miles at a time before being bled and it was on this basis that RT 1 limped into Victoria to waiting guests. One such guest commented that he well remembered RT 1 exhibiting the same nervousness on the occasion of its first public appearance on 13th July 1939 when fuel starvation was also experienced.

With some assistance from others Tim Nicholson threw caution, and the main fuel filter, to the wind and with former Chiswick driving instructor John Killick at the wheel, dressed in period white coat, RT 1 headed finally and uneventfully to Barking sufficiently late as to sweep down the line of RT vehicles waiting for the cavalcade to take her place at the front, much to the astonishment of most of the enthusiasts present.

RT 1 joined Prince Marshall's Obsolete Fleet and was moved from LPC Coachworks to Nunhead garage. Prince died in 1981 but two years later RT 1's restoration had allowed a Certificate of Fitness to be gained and it was able to enter revenue-earning service. It was intended to use RT 1 on route 100 serving the LT Museum and for this role it gained white mudguards, window netting and restricted blinds to give the appearance of wartime condition.

Unfortunately it ran in revenue-earning service for only a short time. Complications arising from Prince Marshall's estate caused Obsolete Fleet to close its operations in late 1983 and route 100 ceased altogether. As the collection was disbanded, RT 1 was sold to Bevan Funnell Ltd and sent to Highpoint, North Carolina in the United States. This seemed, to some, to be such a degrading demise for this historic vehicle that a successful campaign was launched by Michael Dryhurst which resulted in the bus being returned to the United Kingdom in September 1986. The vehicle is now owned by the RT 1 Group which was formed in September 1988 to preserve and maintain it and now carries the livery believed to have been worn when it went into service on route 22 in 1939.

CHAPTER TEN

2RT2 Maintenance

A fascinating manual prepared in three parts by London Transport's Chief Engineer's Department, gave maintenance instructions for those who were to look after the 2RT2 buses, and explains the new type's design and much of the reasoning behind it. As well as the obvious major innovations, there were numerous minor changes in design from previous types based on service experience, showing how the close relationship between the LPTB engineers at Chiswick and those of AEC at Southall benefited the end result.

Quite apart from conveying how deeply thought-out was the design of the whole vehicle, it gives a fascinating glimpse into the atmosphere of London Transport's engineering organisation of the time, with its search for an optimum combination of qualities in the interests of passengers, crew and overall efficiency, particularly in long trouble-free service and ease of maintenance. This in turn conveys a clear sense of purpose and self-confidence – LT knew what it wanted to achieve and how to get it.

Detailed instructions are given for any tasks where the procedure was likely to be in any way unfamiliar, making it clear that close attention to this had been given from the beginnings of the work on the new design. The degree of detail makes it seem probable that quite extensive trial maintenance exercises had been carried out, since at the time no vehicle (not even RT 1) could have built up enough mileage to require some of the work described in great detail.

The main section, simply headed 'Maintenance Bulletin No.24 – The 1939 RT 56-seater double-deck vehicle', running to 40 duplicated typed pages plus five of drawings and diagrams, is not dated, but the description relates to the type with specification soon after being placed in service during the early months of 1940. The second section, similarly headed but with the sub-title 'RT Lubrication' with eight pages plus four of diagrams, is dated 9th March 1940. The third part, 'RT Air Pressure Section – Brakes – Equipment – Gearbox – Indicator', which runs to 18 pages with three of drawings and a diagram, is dated 6th June 1940.

It is noteworthy that the type – the text refers specifically to its code as 2RT2 – is described as 'The 1939 RT' despite the fact that none entered service until 1940, doubtless later than intended because of the outbreak of war.

The opening sentence, headed 'Introduction', is wonderfully understated – 'This vehicle is an entirely new type incorporating several important improvements in both the body and chassis.' Although many items in the general description that follows that statement are covered more fully in the main text of this book, some are significant in the emphasis put upon them and worthy of mention in this section.

Interestingly, the list of salient features begins with the body – 'of composite wood and metal construction with an all-metal bulkhead fitted with perforated panels for sound absorption.' Understandably, the comparisons tend to be with what had gone before, usually the standard STL. Both the conductor's platform and the cab are described as 'larger', which was no doubt true even if the increases were relatively slight and achieved largely by shaping their extremities to take full advantage of the vehicle's overall length.

Reference is made to 'a more forward position for the driver', which again was so, though only by about 2-3in, largely related to the more upright angle of the steering column, giving room for a fully adjustable driver's seat, though of course the lower bonnet line, at almost exactly the level of the seat cushion, gave the RT driver a remarkable line of vision extending down almost to the front edge of the nearside mudguard, ideal for city driving.

Incidentally, this wing is described as being of rubber, indicating that this feature was regarded as standard – it was probably wartime supply difficulties which caused it to apply only to some vehicles. The single headlamp on the nearside and the P-type pass-lamp on the offside dumb-iron are also mentioned as part of the specification, though it is clear that the reference to the latter had been added after the main text had been typed. The cab was claimed to be draught-proof, by virtue of the sealing round the floor-plate and of the pedals, as well as the sliding cab door, all certainly an advance on previous practice – a refinement in the door was the quick-action lift handle for the signalling window, which may have been copied from a type fitted on Rover cars of the period.

On the chassis, the description begins with the engine, and after mentioning its new design, with pot-type direct injection and axial rubber mountings, refers to the large oil cooler and filter below the front of the sump. This last was a feature that

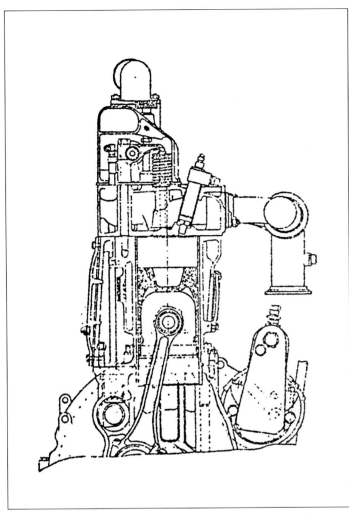

Among the illustrations in the maintenance manual was this cross-section through one of the cylinders of the A185 engine in original form. It is of interest in showing the piston with its cavity, shaped like the inside of a flower-pot, a feature standard on contemporary Leyland engines. Because of this item being of Leyland design, made under licence, AEC did not publicise the internal form of its pot-cavity A180 and A185 engines despite them being built in substantial numbers, mainly for London Transport, in 1938-40.

of mounting. In general, outside London, offside tanks were more usual, though there were some exceptions to that rule.

The fuel tank capacity on RT 19 was later increased to 35 gallons, which became the post-war RT standard. Conversely, RT 2 was for a time fitted with a 22 gallon tank.

In regard to the air system, the point was made that the brake and gear-change pedals operated in opposite senses, the brake pedal allowing air pressure to be admitted to the wheel cylinders, while pressing the gear pedal exhausted air pressure from the gearbox cylinder temporarily until the pedal was released, when air was re-admitted to engage the next gear. The oil bath provided for the handbrake is mentioned, so clearly the later deletion of this feature had not been decided upon at that stage.

The use of shock absorbers at the front but not at the rear is specifically mentioned, though it is pointed out that provision was made for both the addition of them and 'a torsion bar stabiliser' at the rear. The front shock absorbers were telescopic units, made by Newton & Bennett, but it seems that it was decided that even these were unnecessary in London conditions by the time the post-war 3RT type was in volume production.

In describing the exhaust system, the silencer is quoted as being a Blundell acoustical unit. This make was also favoured by Daimler at about this time and, in both cases, quite a 'soft' and unobtrusive exhaust note was evident.

The differential casing was made of Elektron, a light alloy quite widely used in the pre-war period, when weight limits were tight, though it also had the benefit in this application of reducing unsprung weight and hence marginally improving ride comfort. The engine sump was also Elektron and the crankcase in RR 50, an aluminium alloy. This latter had been developed by Rolls-Royce as a suitable material for crankcases where rigidity as well as lightness were important, particularly on aero engines, and was soon to become unavailable for motor vehicles when war production became paramount.

The engine section confirms that it was set to give 100bhp at 1,800rpm, and also that the nominal idling speed was 400rpm (350rpm in gear) but the text goes on to describe how the latter was to be set to give 350rpm in gear as precisely as possible, though the actual adjustment was to be done in neutral to avoid overheating the fluid flywheel. The resemblance to Leyland practice in the combustion system is underlined by the reference to the injectors 'of the long stem type similar to those used on the 10T, the STD and the TF engines'.

The water pump was gear-driven from the timing gears at the front of the engine and, in turn, provided the drive for the chassis-mounted dynamo and compressor via a universally-jointed shaft. The manual does not refer to the later changes which led to the dynamo and compressor being belt-driven from the gearbox, but the gear-driven water pump remained a distinguishing feature of the A185 engine.

had been found on Leyland engines for some time, and may be another example of regard for Leyland engine design features at Chiswick, evident in the choice of combustion system. A noteworthy point was the mounting of the fuel feed pump on the gearbox, from which it was driven, which means that it would cease working when the vehicle was idling in gear – however, it delivered the fuel to the 1¾ gallon reservoir tank under the bonnet which would allow even for quite long periods of this. In fact, when engineers had to attend breakdowns on the road, filling of this tank enabled the bus to travel 10 miles or more. This also applied to all later members of the RT family.

The main fuel tank, of rather modest 27-gallon capacity, was mounted 'as usual on the nearside frame member'. This comment seems slightly surprising at that date, as the STL and other then-recent types had offside tanks, and the nearside position had last been seen on a standard London type on the ST of 1930-31 – possibly the reference was more to the method

It is pointed out that the gearbox, of type D140, 'differs a great deal' from the type fitted to the TF, which had introduced air operation but had an external cylinder, in effect working in the same sense as the driver's foot, and hence used basically the standard D132 gearbox as used on the STL. However, the RT unit's 'internal running gear and associated parts 'are identical with the D132 gear[box] as fitted to the 15STL16 chassis', this latter being the version of the type with flexible engine mountings with numbers STL 2516-2627 which had entered service in 1939 – the engine mounting design of the RT was also similar in principle to that of these latter, the rear unit being identical.

The front axle was new and designed to aid the lightness of steering action by arranging for the centre line of the king pin, when extended to ground level, to almost coincide with the centre of the contact area of the tyre on the road – certainly AEC generally achieved lighter and smoother steering action than most of its competitors in those days long before power assistance became accepted. A noteworthy detail is the statement that 'Wheel discs are fitted to both front and rear wheels.' Clearly this was the intention, but only the 'victory parade' buses, RTs 4 and 39, and RT 97 when adapted for Green Line service, received the front discs intended, though the rear ones were standard in later years.

The section on lubrication was mainly concerned with the RP automatic lubricator, though it also lists the requirements of the chassis generally. The RP unit fed 24 points judged to justify frequent supply. It was operated by the air system and one shot of oil was fed each time the brake pedal was operated, the mechanism moving on to another point at the next brake application. There was considerable interest in such devices at that time, not uncommon on the more expensive private cars but the RP unit, with its positive feed at a pressure of over 1,000lb/sq in of very small quantities of oil to each point in turn, was more successful than most. There remained quite a number of other points requiring attention with oil or grease gun, some where automatic feed was impractical or where the need was infrequent.

The section on the air pressure system begins with the statement that 'It is practically the same as used on the TF vehicles....', which was so in principle, though a list of the 2RT's differences includes some fairly substantial points. It would have been mainly of value within Chiswick works, since operation of both types from one garage must have seemed unlikely even at the beginning, save perhaps for the private hire TF operation from Victoria (GM). The gearbox differences already mentioned were the most obvious points, there also being a separate tank for the RT gearbox, but the front and rear brake operating cylinders also differed because of chassis design differences. A test panel developed for the TF system could also be used for the RT with a special adaptor.

At that stage, air-pressure braking was virtually unknown among British motor vehicles, though it had become common on trolleybuses, so maintenance staff were dealing with unfamiliar equipment. It was explained that there were three interconnected airtight sections – the compression and storage section, the brake operating section, and the gearbox section. Nowadays, air pressure braking is standard for most heavy vehicles, so much of the detail comment would be familiar but particular interest attaches to the air compressor section.

The action of the rotary compressor is described, it being based on a principle of one cylinder rotating eccentrically within a larger one, the air to be compressed being trapped by a plate, the edge of which was pressed against the rotating cylinder with ports at the ends to admit air and allow it to pass to the system. It was an ingenious design, but it is noteworthy that 'if excessive oil leakage, noise or persistent overheating' occurred it was advised that the unit was to be changed and sent to Chiswick for examination or repair. A gauge was provided in the system below the cab floor plate for use by garage staff.

Compressor build-up could be timed with the engine run at idling speed to provide a basis of comparison and minimum performance, but it is significant that the bulletin stated that 'No figures can be given, but providing the build-up commences at once and reaches 80 lbs within a few minutes, it may be considered satisfactory.' This suggests that individual units even in good order were apt to be inconsistent in their performance, and in practice considerable trouble was experienced, leading to replacement with a more conventional reciprocating compressor with better and more reliable performance. Another note states that 'Sluggish braking may be caused by oil pockets existing in the pipe line of the operating side, and should be looked for at U-bends or at low hanging pipe loops into which oil can collect', which suggests that though an oil separator was incorporated it may not have been very effective.

A later note describes how pressure gauges could be attached at various points in the system, giving more specific times for performance. A test board with three gauges, as already made up for the TF, could be used for checking performance on a moving vehicle.

All in all, a picture emerges of considerable time and judgement being needed to check the performance of the air system, and clearly both the design and maintenance was still at a learning-curve stage. Fortunately, London Transport was given an opportunity to rectify the situation by the cessation of further deliveries of RT buses due to the war. Even so, it is noteworthy that despite all the problems and distractions of the time, London Transport's engineers maintained the will to persevere and succeed. This was to benefit the whole commercial vehicle operating and manufacturing industry, for ultimately air brakes were to become standard practice on most heavy-duty vehicles, as still applies today.

An insight to methods for the work done at the garages has been provided by Alan Pearce, who was one of the maintenance team at Chelverton Road and Putney Bridge, the two garages most associated with the type. The following is based quite closely on his notes of the methods in use during the 1950-53 period, the final years of 2RT2 operation in central London. A system of regular maintenance inspections of the vehicles was in force, and they were referred to simply as 'rotas'.

The 'A' Rota was four-weekly and basically consisted of a good look round each vehicle, both mechanically and in regard to the body. Generally two mechanics inspected four or five vehicles per day. The 'B', 'C' and 'D' Rotas were carried out by a team described as 'The Dock', referring to the inspection pits where such work was carried out. The 'A' Rota men were not included in this group.

The 'B' Rota took place every 12 weeks and was more thorough than the 'A', involving oil changes and partial strip down of selected mechanical components. The body and electrical side were also included, parts requiring replacement being dealt with by men of the appropriate trades. The 'C' Rota, at 24 weeks, was on a 'strip, examine and renew as required' basis, dealing with some additional items.

The 'D' Rota, with a 48-week interval, was more of an overhaul, where most mechanical and electrical units were removed or stripped to bare shells. Engines and gearboxes had only selected parts removed at this time, though if it was discovered during the examination that these units needed to come out for more major attention, then this work was deferred until the following day and carried out by the 'miscellaneous men', exchange units being supplied from Chiswick. Bodywork was not stripped out to the same degree and any serious defects were dealt with as a 'works' job at the major overhaul; any serious accident damage also led to the bus being sent to Chiswick for repair.

The mechanics on 'The Dock' at Chelverton Road were divided into six two-man teams, each dealing with two buses daily. They specialised, so the 'Fronts' dealt with the front axle, springs and front cardan shaft; the 'Rears' dealt with the rear axle including the differential unit, springs, and the rear cardan shaft, and the 'Engines' men dealt with the radiator, engine and components thereon including the fluid flywheel. In addition, a bench fitter was in attendance on the gearbox internals, covering four buses.

Most of Alan Pearce's time was on an 'Engines' team, it being noteworthy that one of his partners, Cyril Nash, had served his time on the South Eastern & Chatham Railway. Work started at 7.30am, when the radiator would be dropped and the fuel pump and injectors removed, degreased and taken to the pump shop for testing, and the engine oil drained. A quick break for breakfast at the cafe next door was followed by removal of the rocker-box cover to check the valve gear and all filters before checking the fluid flywheel. If only one thread on the gland housing was showing, the gland and housing came down for renewal, but not before the general condition of the whole unit had been checked. In practice, the spigot race never seemed to last more than a year, leading to wear on the main race, so the complete fluid flywheel was normally changed at a 'D' dock, the gland frequently so at a 'C' dock. Fuel pipe runs and coolant hoses were checked, the latter normally changed as a matter of course. Then reassembly began and, usually just after lunch, the engine block and radiator were flushed and a yellow anti-corrosion additive put into the cooling water. By 2pm, the engine should have been run up and all adjustments carried out.

The Dock Foreman liked to be notified before he had his cup of tea at 2.15 that all buses were ready for test, a job he shared with the Running Shift Foreman. Buses turned right into Putney High Street and climbed Putney Hill, giving an opportunity to make good use of the gearbox, and then went via the A3, with a good 'blat' downhill towards Putney Vale with much applying of foot and hand brakes. Any defects, almost always poor or binding brakes, were corrected and, if then satisfactory, the bus was released for service.

Appendix One:
Standard Specification of the 2RT2 Class, as Built

Chassis (AEC, Southall)

Engine
AEC type A185
Six-cylinder direct-injection with pot-cavity combustion system.
Bore and stroke 120mm by 142mm. Swept volume 9.6 litres.
Power output (standard rating) 100 bhp at 1,800 rpm.
Gear driven timing including drive to water pump, drive via flexibly jointed shaft to frame-mounted compressor and dynamo.
Belt friven fan.
Engine flexibly mounted.

Transmission
Fluid flywheel – 16in diameter, AEC unit J150.
Gearbox – Wilson-type preselective epicyclic, with air pressure operation, AEC unit D 140 (except RT 147-151), mounted amidships in chassis, preselector control lever mounted on steering column.
Gearbox ratios: Top 1:1; Third 1.64:1; Second 2.53:1; First 4.51:1; Reverse 6.9:1.
Cardan shafts connecting fluid flywheel to gearbox and gearbox to rear axle with Hardy Spicer universal joints.

Rear Axle
Underslung worm drive, fully floating, ratio 5.75:1, AEC unit F187.

Brakes and Air System
Footbrake operated through compressed-air system, front cylinders mounted above king pins, rear cylinders on frame, acting through linkage to rear axle brakes.
Drums 16$^{3}/_{4}$in diameter with R.P. triple-pawl automatic brake adjusters at front and rear. Frame mounted rotary compressor driven by flexibly jointed shaft from engine, supplying air unloader valve to duplex main pressure tank with separate circuits for brakes, gear change system and R.P. automatic lubricator.
Handbrake ratchet gear and quadrant encased in oil bath.

Electrical System
24 volt, two 12 volt accumulators fitted in compartment below staircase, booster socket located on the offside below driver's step.

Chassis Frame
Pressed-steel sidemembers with maximum depth of 11$^{1}/_{4}$in and 3in flanges, supports for rear platform formed as part of main pressing.
Most cross-members tubular, but forged front cross-member incorporating front engine mounting point, tubular second member supports rear engine mounting and the support bracket for the compressor and dynamo. Fuel tank (27 gallons) mounted nearside.

Front Axle and Steering
Beam front axle, king pins angled to give high degree of 'centre point'. Steering gear AEC worm and nut type.

Suspension, wheels and tyres
Leaf springs, front 4ft 2in long, nine leaves; rear 5ft 2in long, ten leaves. Newton and Bennett C2-type telescopic shock absorbers at front axle only. Front tyres 36x8 high pressure, rear tyres 9.00x20 low pressure twin.

Body (London Passenger Transport Board, Chiswick)
Double-deck, rear entrance, 56 seats, 30 in upper deck, 26 in lower saloon. Composite wood and metal front bulkhead; double skinned upper deck roof. Simplastic glazed windows; some with winding half drop sections. Sliding cab door fitted with quick action signalling window. Route number boxes at roof level front and rear.
One headlamp fitted to nearside. Wings and mudguards manufactured in rubber, alternatives in steel fitted to some vehicles.
Mudflap fitted below rear platform.

Initial Livery
Exterior: Red, broken white on centre band and widow surrounds (radiused ends on lower deck); grey roof (rear dome, surround of front roof route number box and lining above upper deck windows in red), black wings and mudguards, india red wheel centres.

Interior: Cream painted ceilings; cream and green Rexines applied to window shrouds, brown Rexine applied to panels below windows and staircase area. Lower compass panels and lower front bulkhead covered in linoleum; cork tiled floor, wooden slatted gangways and platform. Staircase treads and risers, conductor's locker area and flywheel cover painted brown; non-polished sections of seat frames painted green. Seat backs and squabs trimmed in moquette. Cab: brown painted below window line; matt black window edges and ceiling.

Main Dimensions
Overall length: 25ft 11$^{3}/_{8}$in.
Overall height: 14ft 3$^{1}/_{6}$in laden.
Overall width: 7ft 6in maximum; (7ft 4in over pillars)
Wheelbase: 16ft 4in.
Front overhang: 2ft 1$^{3}/_{8}$in.
Bonnet length (front of radiator to bulkhead): 4ft 6$^{1}/_{4}$in.
Front track: 6ft 3$^{9}/_{16}$in at wheel centres
Rear track: 5ft 9$^{1}/_{2}$in.

Axle Ratios
Rear axle ratios, all originally 5$^{3}/_{4}$:1 were changed to 5$^{1}/_{5}$:1 except where otherwise stated, and although no dates are recorded, it is understood that this occurred after the engines had been converted to toroidal.
Three vehicles, RT 39, 44 and 56 temporarily converted to 5$^{3}/_{4}$:1 before being again changed to 5$^{1}/_{5}$:1.
The record cards for RT 16, 22, 33, 43, 45, 47, 49, 59, 68, 74, 76, 85, 88, 92, 94, 96, 97, 103, 110, 111, 113, 116, 120, 125, 130, 131, 138, 140, 142, 143 and 148-150 show no change from 5$^{3}/_{4}$:1.
The cards for RT 11, 14, 28, 37, 89, 100 and 134 show a change from the 5$^{3}/_{4}$:1 to 5$^{1}/_{6}$:1, the latter being AEC's post-war equivalent to 5$^{1}/_{5}$, usually associated with a redesign of the rear axle. In the special case of RT 19, the ratio changed from the original to 5$^{1}/_{5}$:1 when this vehicle's engine was converted to toroidal in 1940, and then to 5$^{1}/_{6}$:1 post-war. A later case of similar sequence was RT 12.
The record card for RT 23 shows a change from 5$^{3}/_{4}$:1 ratio to 5$^{2}/_{5}$:1 before reverting to 5$^{3}/_{4}$:1 and then finally 5$^{1}/_{5}$:1. This is somewhat surprising as 5$^{2}/_{5}$, equivalent to 5.4:1 in decimals, was not a standard AEC axle ratio for either the pre-war or post-war Regent ranges, though it was a standard Leyland ratio. Cutting a special worm and wormwheel to give this ratio would have been quite expensive and it seems possible that this entry could be an error.

Appendix Two: 2RT2 Body Numbers and Dates Vehicles taken into Stock

Body No.	Bonnet No.	Into Stock	Body No.	Bonnet No.	Into Stock	Body No.	Bonnet No.	Into Stock
280	RT 15	28.12.39	330	RT 50	19.3.40	380	RT 139	22.4.40
281	RT 19	28.12.39	331	RT 34	16.2.40	381	RT 104	23.4.40
282	RT 16	28.12.39	332	RT 21	16.2.40	382	RT 119	26.4.40
283	RT 20	28.12.39	333	RT 99	19.2.40	383	RT 121	27.4.40
284	RT 39	29.12.39	334	RT 36	19.2.40	384	RT 116	29.4.40
285	RT 43	29.12.39	335	RT 13	21.2.40	385	RT 92	30.4.40
286	RT 44	29.12.39	336	RT 14	24.2.40	386	RT 125	3.5.40
287	RT 38	30.12.39	337	RT 10	24.2.40	387	RT 129	1.5.40
288	RT 47	30.12.39	338	RT 94	25.2.40	388	RT 88	2.5.40
289	RT 37	30.12.39	339	RT 8	25.2.40	389	RT 136	4.4.40
290	RT 56	1.1.40	340	RT 67	28.2.40	390	RT 126	6.5.40
291	RT 30	1.1.40	341	RT 110	28.2.40	391	RT 131	7.5.40
292	RT 49	1.1.40	342	RT 64	29.2.40	392	RT 62	3.4.40
293	RT 25	1.1.40	343	RT 17	4.3.40	393	RT 128	10.5.40
294	RT 5	1.1.40	344	RT 108	6.3.40	394	RT 127	5.4.40
295	RT 48	12.1.40	345	RT 84	6.3.40	395	RT 137	11.5.40
296	RT 54	12.1.40	346	RT 115	7.3.40	396	RT 93	14.5.40
297	RT 40	12.1.40	347	RT 113	7.3.40	397	RT 70	15.5.40
298	RT 57	12.1.40	348	RT 73	8.3.40	398	RT 82	16.5.40
299	RT 52	12.1.40	349	RT 98	11.3.40	399	RT 118	21.5.40
300	RT 53	12.1.40	350	RT 11	13.3.40	400	RT 138	22.5.40
301	RT 31	12.1.40	351	RT 90	15.3.40	401	RT 141	24.5.40
302	RT 28	12.1.40	352	RT 79	15.3.40	402	RT 75	25.5.40
303	RT 59	17.1.40	353	RT 74	18.3.40	403	RT 80	28.5.40
304	RT 35	17.1.40	354	RT 96	19.3.40	404	RT 76	29.5.40
305	RT 3	18.1.40	355	RT 33	21.3.40	405	RT 106	30.5.40
306	RT 42	18.1.40	356	RT 102	26.3.40	406	RT 69	30.5.40
307	RT 29	18.1.40	357	RT 87	27.3.40	407	RT 71	4.6.40
308	RT 6	20.1.40	358	RT 112	28.3.40	408	RT 91	31.5.40
309	RT 41	20.1.40	359	RT 45	29.3.40	409	RT 101	3.6.40
310	RT 27	22.1.40	360	RT 109	1.4.40	410	RT 111	6.6.40
311	RT 60	22.1.40	361	RT 58	1.4.40	411	RT 132	8.6.40
312	RT 61	24.1.40	362	RT 95	2.4.40	412	RT 123	10.6.40
313	RT 32	24.1.40	363	RT 89	6.8.40	413	RT 114	11.6.40
314	RT 22	25.1.40	364	RT 107	4.4.40	414	RT 130	12.6.40
315	RT 63	25.1.40	365	RT 81	16.4.40	415	RT 145	13.6.40
316	RT 4	27.1.40	366	RT 97	29.4.40	416	RT 140	14.6.40
317	RT 2	29.1.40	367	RT 85	6.4.40	417	RT 134	17.6.40
318	RT 23	29.1.40	368	RT 105	9.4.40	418	RT 143	17.6.40
319	RT 18	29.1.40	369	RT 117	10.4.40	419	RT 144	18.6.40
320	RT 7	6.2.40	370	RT 55	11.4.40	420	RT 120	20.6.40
321	RT 9	6.2.40	371	RT 103	11.4.40	421	RT 146	21.6.40
322	RT 66	6.2.40	372	RT 65	12.4.40	422	RT 135	24.6.40
323	RT 12	6.2.40	373	RT 77	15.4.40	423	RT 142	26.6.40
324	RT 46	8.2.40	374	RT 78	16.4.40	424	RT 100	28.6.40
325	RT 51	10.2.40	375	RT 133	17.4.40	425	RT 147	1.3.41
326	RT 26	12.2.40	376	RT 124	18.4.40	426	RT 149	18.8.41
327	RT 72	13.2.40	377	RT 83	19.4.40	427	RT 151	14.1.42
328	RT 24	12.2.40	378	RT 122	22.4.40	428	RT 150	27.5.41
329	RT 68	12.2.40	379	RT 86	22.4.40	429	RT 148	16.10.41

A total of 32 2RT2 buses received different bodies during the first few years of their existence, in a series of exchanges which ran, intermittently, from March 1941 until March 1947. In most cases, vehicles received their 'new' bodies during overhaul. However, in the final year of the War, the Enemy's V1 Flying Bomb campaign accounted for serious damage to five members of the class, and as a result all were sent to Chiswick Works where three received refurbished bodies in a sequence of intricate moves which also included the prototype and RT 19.

The first exchange involved the swapping of the bodies belonging to RT 18 and RT 29 in March 1941, when both vehicles were in store at the London Terminal Coach Station. For some unknown reason the bodies were reunited with their original chassis and this action may have been purely experimental to ascertain that such an operation was possible within the confines of a unit not primarily constructed for such work.

The first permanent exchange of bodies did not occur until April 1943, the vehicles selected on this occasion being RT 9 and RT 49 whilst both were undergoing their first overhaul at Chiswick. At this time the main programme to replace the compressors of 49 members of the class had barely begun and of the two vehicles only RT 9 had been identified to receive the modification. The conversion work also involved alterations to the body and it was that from RT 49 which was modified, RT 9's body remaining in its initial condition for fitting to RT 49's chassis in order to complete the exercise. RT 49 then soldiered on, as a standard 2RT2, until the autumn of 1949 when it was converted to the 3/2RT2/2 specification.

RTs 22, 31 and 67 were the subject of the next interchange of bodies which occurred as the overhaul of each of these vehicles was being completed in late January 1944. This time all three vehicles were converted to the 1/2RT2/1 specification and therefore their bodies and chassis were compatible. In the event the chassis from RT 67 received the body from RT 31, whose own chassis was equipped with the body from RT 22, with the fitting of the body from RT 67 to RT 22's chassis completing the three-way manoeuvre. When, on 2nd January 1951, RT 22 unceremoniously overturned on Wimbledon Hill and was subsequently written off, it was the body which was initially fitted to RT 67 that was lost; the body fitted to RT 22, when new, surviving for at least another eight and a half years.

January 1944 also saw RT 66 and RT 52 emerge from overhaul having been modified to the 1/2RT2/1 classification. Like the majority of the 2RT2s passing through Chiswick at this time, these buses retained their original bodies but five months later both vehicles were back at the West London Works in company with RT 87, the buses having sustained serious damage in flying bomb incidents. The action taken by Chiswick resulted in the bodies and chassis from these vehicles being permanently parted, the body from RT 66 being scrapped in the process. V1 explosions also accounted for the major damage to RT 59 and RT 97 but, as recorded in the main text, both vehicles were individually repaired and as a result retained the bodies which had been fitted at the time of their construction. A chronological listing of the course of events involving RTs 52, 66 and 87 in order to return these vehicles to service is included in the list opposite.

There was one more exchange of bodies in 1944. It occurred during the overhauls of RT 138 and RT 144 during April, from which both emerged as unconverted 2RT2 buses.

It was to be some months before the next body changes were attempted and this time nineteen vehicles were involved in a series of displacements, initiated by the arrival of RT 85 at Chiswick for its second overhaul in July 1946. Here the vehicle remained as a chassis for the next eight months following the removal of its body which was overhauled and fitted to RT 43. The programme which had commenced in March 1943, to modify 49 members of the class to the 1/2RT2/1 specification, had been completed some twenty months before RT 85 entered Chiswick and therefore vehicles which had been amongst the first converted were now due for second overhauls.

The fact that there were two types of body in existence did not affect the next displacement when the body from RT 43 was fitted to the chassis of RT 94. But when the body from RT 94 was identified for mounting on RT 46, the situation arose of the pairing of a converted chassis with an unconverted body. Chiswick had already been aware of the position and had singled out RT 23, itself converted to 1/2RT2/1, to exchange bodies with RT 46 in a completely separate operation. This work even progressed as far as uniting the body from RT 46 with the chassis from RT 23 and the body from RT 23 with the body from RT 46. However for no recorded reason the latter pairing was aborted, after ten days had elapsed, when the body originally fitted to RT 23 was removed and mounted on the unconverted chassis belonging to RT 24. This action, together with the fitting of the body from RT 94 to RT 46, created two sub groups (although both had previously been created during the wartime repairs). RT 46 therefore became 1/2RT2 (body not converted) and RT 24 became 2RT2/1 (chassis not converted). In the event RT 46's replacement body was eventually altered and the vehicle became 1/2RT2/1 once more. However RT 24 retained its modified body until the vehicle was altered to the 3/2RT2/2 specification in late autumn 1949.

The rest of this sequence of body displacements continued without any further problems, the remaining vehicles involved being standard 2RT2s. The final moves came in the spring of 1947 when the body from RT 37 was fitted to RT 56, whose own body was fitted after overhaul to RT 85's chassis, this being the last ever transfer of a 2RT2 body to another chassis.

Initial Experimental Body Exchange

17.3.41	Body No. 319 (ex-RT 18) fitted to RT 29 chassis.
17.3.41	Body No. 307 (ex-RT 29) fitted to RT 18 chassis.

Note: Both bodies were reunited with their original chassis before the vehicles re-entered service.

First Exchanges During Overhaul

4.43	Body No. 292 (ex-RT 49) fitted to RT 9 chassis, vehicle converted to 1/2RT2/1.
4.43	Body No. 321 (ex-RT 9) fitted to RT 49 chassis, vehicle later converted to 3/2RT2/2 specification (9/49 - 10/49).
24.1.44	Body No. 301 (ex-RT 31) fitted to RT 67 chassis, vehicle classified 1/2RT2/1.
26.1.44	Body No. 340 (ex-RT 67) fitted to RT 22 chassis, vehicle classified 1/2RT2/1.
28.1.44	Body No. 314 (ex-RT 22) fitted to RT 31 chassis, vehicle classified 1/2RT2/1.
29.4.44	Body No. 400 (ex-RT 138) fitted to RT 144 chassis. Body No. 419 (ex-RT 144) fitted to RT 138 chassis.

Dealing with War Damaged Vehicles

14.1.44	RT 66 released from overhaul, with original body (No. 322), vehicle classified 1/2RT2/1.
20.4.44	RT 52 released from overhaul, with original body (No. 299), vehicle classified 1/2RT2/1.
13.6.44	Start of the Enemy's V1 flying bomb campaign.
6.44	RT 52 withdrawn as a result of Enemy action.
17.6.44	RT 87 withdrawn as a result of a flying bomb incident at Clapham Common whilst in service on Route 37 (AF).
20.6.44 - 24.6.44	RT 87 overhauled and fitted with Body No. 299 (ex-RT 52) on 22.6.44: vehicle should have been classified 2RT2/1 (not recorded). RT 87 finally converted to 3/2RT2/2 specification (12.50 - 2.51)
22.6.44	Body No. 357 (ex-RT 87) temporarily mounted on RT 52, vehicle should have been classified 1/2RT2 (not recorded)
26.6.44	Body No. 357 (ex-RT 87) removed from RT 52
26.6.44	RT 52 to AEC (chassis only) for attention following Enemy action.
30.6.44	RT 66 withdrawn and sent to Chiswick as the result of Enemy action.

1.7.44	RT 87 licensed for service at Chelverton Road Garage.
20.7.44	Body No. 322 (RT 66) removed from chassis and scrapped.
6.9.44	Body No. 357 (ex-RT 87) fitted to RT 66, vehicle initially classified 1/2RT2, later 1/2RT2/1.
1.10.44	RT 66 licensed for service at Putney Bridge Garage.
19.3.44	RT 19 withdrawn and sent to Chiswick for overhaul.
23.3.45	RT 52 chassis returned to Chiswick from AEC.
26.3.45	Body No. 281 removed from RT 19 chassis. Chassis to AEC.
27.3.45	Body No. 281 (ex-RT 19) fitted to RT 52, vehicle classified 1/2RT2: later 1/2RT2/1.
3.5.45	RT 52 overhauled at Chiswick.
4.5.45	RT 52 licensed for service at Putney Bridge Garage.
3.8.45	RT 19 chassis returned to Chiswick from AEC.
18.8.45	RT 1 withdrawn at Chiswick.
29.9.45	RT 19 chassis to AEC.
27.11.45	RT 19 chassis returned to Chiswick from AEC; Body No. 18246 (ex-RT 1) fitted, vehicle classified 2/2RT1.
1.3.46	RT 19 licensed for service at Putney Bridge Garage.
4.9.46	RT 1 chassis dismantled.

The Final Body Exchanges

Date Fitted	Stock No.	Body No.	Classification
1.8.46	RT 43	367 (ex-RT 85)	2RT2
14.8.46	RT 94	285 (ex-RT 43)	2RT2
21.8.46	RT 46	338 (ex-RT 94)	1/2RT2
9.7.46	RT 23	324 (ex-RT 46)	1/2RT2/1
18.7.46	RT 24	318 (ex-RT 23)	2RT2/1
24.7.46	RT 48	328 (ex-RT 24)	2RT2
31.7.46	RT 68	295 (ex-RT 48)	2RT2
14.8.46	RT 10	329 (ex-RT 68)	2RT2
20.8.46	RT 11	337 (ex-RT 10)	2RT2
30.8.46	RT 108	350 (ex-RT 11)	2RT2
9.9.46	RT 117	344 (ex-RT 108)	2RT2
26.9.46	RT 30	369 (ex-RT 117)	2RT2
28.9.46	RT 16	291 (ex-RT 30)	2RT2
15.10.46	RT 20	282 (ex-RT 16)	2RT2
4.11.46	RT 5	283 (ex-RT 20)	2RT2
2.12.46	RT 47	294 (ex-RT 5)	2RT2
19.12.46	RT 37	288 (ex-RT 47)	2RT2
14.1.47	RT 56	289 (ex-RT 37)	2RT2
14.3.47	RT 85	290 (ex-RT 56)	2RT2

Appendix Four:
Codes Used on Early RT Buses

London Transport's system of applying engineering codes to distinguish chassis and body types was introduced soon after it came into existence in 1933, no doubt as an aid to keeping track of the variety then found in the fleet. The idea, based on the type letter system inherited from the LGOC, was to provide a simple means of identifying differences, with particular emphasis on the compatibility of types of body and chassis.

Chassis code numbers preceded the vehicle type letters, the prototype RT chassis thus being coded 1RT, and body code numbers followed the letters, and hence its body was type RT1, the complete vehicle being identified by the combined code 1RT1. Similarly the production batch had chassis type 2RT and body type RT2 as built, hence 2RT2 complete. In ordinary conversation among London Transport staff, there was a tendency to use parts of these codes, not always in accordance with the above principles, so RT2, or even just '2' was sometimes intended to convey the complete 2RT2 type of vehicle.

Minor variants were denoted by what were called 'stroke' numbers applied before the main chassis type number or after the main body number, creating numbers such as 1/2RT2/1, as indicated below.

In general, the system worked well, enabling vehicle types to be indicated quite comprehensively and succinctly. Yet, inevitably it was not always consistent, relying, as did London Transport's basically methodical records as a whole, on the decisions of those who operated the system, sometimes differing in their approach and perhaps not always fully understanding the way it was meant to work, quite apart from being liable to make errors.

A further problem which occurs in several instances in regard to these buses is a discrepancy between what the record cards show and what was observed by reliable witnesses as being displayed by the small brass code plates on the chassis or body of the vehicle. In some cases this may have been simply a clerical omission, but there may have been instances of one part of the organisation not being informed as to what another was doing, especially amid the pressures of getting the fleet back into good order after the 1939-45 war years.

Code	Fleet Nos involved	Description	Date
1RT1	RT 1	Prototype bus, as built (evidently its period as ST 1140 was ignored because of its temporary nature).	1939
2/1RT1	RT 1	Significance not known. Appears on record card of vehicle with no clue as to meaning or date. (1/1RT seems not to have been used.)	?
2RT2	RT 2-151	Initial production version.	1940
1/2RT2/1	RT 143 (converted before entering service 1941), RT 2, 3, 6, 8, 9, 15, 17, 22, 23, 26, 27, 31, 32, 39, 40, 46, 52-4, 57-61, 64-7, 71, 75, 80, 83, 89, 97, 99-101, 103, 106, 110, 121, 122, 125, 130, 139-141, 145 and 146.	Belt-driven reciprocating compressor, driven from gearbox (50 conversions in all, those after RT 143 completed between March 1943 and August 1944).	1941

In one or two cases it seems that the chassis alteration was made and the vehicle returned to service before the corresponding change to the floor traps of the body was done, thus becoming temporarily 1/2RT2.

2/2RT2	RT 19	Reported as applied to RT 19 as at August 1942 overhaul, but not shown on card for this bus (may cover 1940 mods).	1942
2/2RT1	RT 19	RT 19 evidently in form when first converted to act as prototype for for post-war RT, with body from RT 1. (See also under 3RT3).	1945
2/2RT	RT 1?	Appears on chassis photograph of RT1, evidently related to plan to update it before decision to use RT 19.	1945

3/2RT2/2	RT 4, 5, 7, 10-14, 16, 18, 20, 21, 24, 25, 28-30, 33-38, 41-45, 47-51, 52, 55, 56, 62, 63, 68-70, 72-74, 76-79, 81, 82, 84, 86-88, 90-96, 98, 102, 104, 105, 107-109, 111-120, 123, 124, 126-129, 131-138, 142, 144, 147-151	Belt-driven dynamo and reciprocating compressor, both driven from gearbox. (99 conversions).	1948 to 1951

RT 85 is recorded as unaltered from 2RT2 when scrapped in 6/49 after the fire in 5/49, but had been due for conversion.

A further complication here is that RT 97, as extensively converted for PAYB operation in 1945, before any of the above conversions began, has been reported to have shown RT2/2 on its body plate at that stage, even though this was not quoted on its record card. In theory this created a situation when one code meant two quite different things, clearly not intended. In practice, it made no difference, as by the time the code RT2/2 was coming into use on the buses with the compressor and dynamo change listed above, RT 97 was in process of receiving further large-scale modification to become RTC 1, becoming code 5RT5, as indicated below.

3RT1	RT 19	Re-classification following full conversion of chassis to post-war specification.	1948
3RT1/1	RT 19	The body plate on this vehicle was checked as reading RT1/1 in January 1953, but this does not appear on record card. (Chassis plate not legible at the time.)	1953

3RT3	RT 19	This code, the post-war production standard, appears on the card for RT 19 before 2/2RT1, possibly due to misunderstanding when it was chosen as basis for post-war design.	1945(?)
5RT5	RTC 1	The code carried by RTC 1 when it re-entered service after conversion from RT 97. It had originally been planned to make it 9RT10, coming after the SRT conversions, but evidently the unused 5RT5 was preferred.	1949

Appendix Five:
2RT2 Conversions to 1/2RT2/1

Conversion Completed	Stock No.
11/40 (prior to service)	RT 143
3/43	RT 39
3/43	RT 103
3/43	RT 59
3/43	RT 46
4/43	RT 9
4/43	RT 23
5/43	RT 110
6/43	RT 125
7/43	RT 139
9/43	RT 100
12/43	RT 6
12/43	RT 32
12/43	RT 26
12/43	RT 3
1/44	RT 17
1/44	RT 61
1/44	RT 66
1/44	RT 27
1/44	RT 54
1/44	RT 2
1/44	RT 58
1/44	RT 22
1/44	RT 31
2/44	RT 67
2/44	RT 8
2/44	RT 99
2/44	RT 40
2/44	RT 60
2/44	RT 122
2/44	RT 75
2/44	RT 64
3/44	RT 83
3/44	RT 145
3/44	RT 71
3/44	RT 89
3/44	RT 141
3/44	RT 15
3/44	RT 130
3/44	RT 121
3/44	RT 57
3/44	RT 106
3/44	RT 140
4/44	RT 65
4/44	RT 53
4/44	RT 101
4/44	RT 52
4/44	RT 97
4/44	RT 80
11/44 (possibly earlier)	RT 146

Appendix Six:
2RT2 Conversions to 3/2RT2/2 (all at overhaul)

Off	Return	Bonnet No.	Off	Return	Bonnet No.
7/48	9/48	RT 73	3/50	6/50	RT 41
7/48	9/48	RT 74	4/50	5/50	RT 56
7/48	9/48	RT 128	4/50	6/50	RT 38
8/48	10/48	RT 36	5/50	6/50	RT 28
8/48	10/48	RT 62	5/50	6/50	RT 29
8/48	10/48	RT 86	5/50	7/50	RT 18
8/48	10/48	RT 114	6/50	8/50	RT 147
8/48	10/48	RT 123	9/50	10/50	RT 124
8/48	10/48	RT 136	9/50	10/50	RT 138
9/48	11/48	RT 78	9/50	10/50	RT 144
9/48	11/48	RT 79	9/50	11/50	RT 21
9/48	11/48	RT 104	9/50	11/50	RT 51
12/48	1/49	RT 82	10/50	11/50	RT 81
12/48	1/49	RT 88	10/50	11/50	RT 105
12/48	1/49	RT 93	10/50	11/50	RT 118
12/48	1/49	RT 132	10/50	12/50	RT 42
12/48	1/49	RT 134	10/50	12/50	RT 70
12/48	1/49	RT 137	10/50	12/50	RT 84
1/49	3/49	RT 76	10/50	12/50	RT 112
1/49	3/49	RT 77	10/50	12/50	RT 129
1/49	3/49	RT 91	11/50	12/50	RT 14
1/49	3/49	RT 151	11/50	12/50	RT 72
3/49	5/49	RT 135	11/50	12/50	RT 150
3/49	5/49	RT 142	11/50	1/51	RT 90
4/49	5/49	RT 149	12/50	1/51	RT 44
5/49	8/49	RT 111	12/50	2/51	RT 12
5/49	8/49	RT 120	12/50	2/51	RT 34
7/49	8/49	RT 4	12/50	2/51	RT 45
7/49	8/49	RT 7	12/50	2/51	RT 50
7/49	8/49	RT 148	12/50	2/51	RT 63
9/49	10/49	RT 48	12/50	2/51	RT 87
9/49	10/49	RT 49	12/50	2/51	RT 98
9/49	11/49	RT 24	12/50	2/51	RT 127
9/49	11/49	RT 69	12/50	2/51	RT 133
9/49	12/49	RT 16	1/51	2/51	RT 95
10/49	12/49	RT 10	2/51	3/51	RT 96
10/49	12/49	RT 35	3/51	4/51	RT 92
10/49	12/49	RT 43	3/51	5/51	RT 109
10/49	12/49	RT 68	3/51	5/51	RT 115
10/49	12/49	RT 94	4/51	5/51	RT 33
11/49	12/49	RT 30	4/51	5/51	RT 102
11/49	1/50	RT 11	4/51	5/51	RT 107
11/49	1/50	RT 13	4/51	5/51	RT 126
11/49	2/50	RT 108	5/51	6/51	RT 131
12/49	1/50	RT 5	5/51	7/51	RT 116
12/49	2/50	RT 20	5/51	7/51	RT 119
12/49	2/50	RT 117	6/51	8/51	RT 113
1/50	2/50	RT 25	7/51	8/51	RT 55
1/50	2/50	RT 47	9/51	10/51	RT 52
2/50	3/50	RT 37			

Appendix Seven:
Summary of Routes Allocated RTs from 1939 to 1945

This list gives details of routes allocated RTs between the entry into service of RT 1 and the end of 1945. Only those garages allocated the type are shown. In many cases the routes were shared with other garages which used older types for their allocation. Route 37 for example was shared with Mortlake until May 1940, LTs being used.

August 1939	22	AF	RT 1

January 1940	22	AF	One vehicle
	37	AF	

March 1940	28	AF	
	30	AF	
	37	AF	

May 1940	14	F	Partly STL until June 1940
	28	AF	
	30	AF	
	37	AF	
	72	AF	

August 1940	37	AF	Most RTs delicensed

November 1940	37	AF	
	72	AF	

April 1941	14	F	
	37	AF	
	72	AF	

June 1941	14	F	
	37	AF	
	72	AF	
	74	F	Sat & Sun only
	85	F	

October 1941	14	F	
	30	AF	
	37	AF	
	52	GM	
	74	F	
	77A	GM	Partly STL
	85	F	
	93	F	
	96	F	

March 1942	14	F	
	30	AF	
	37	AF	
	52	GM	
	72	AF	
	77A	GM	Partly STL
	85	F	
	93	F	
	96	F	
	137	GM	Sundays between April and October 1942

October 1943	14	F	
	30	AF	
	37	AF	
	72	AF	
	74	F	
	77A	GM	Partly STL
	85	F	
	93	F	
	96	F	

December 1943	14	F	
	30	AF	
	37	AF	
	72	AF	
	74	F	
	77A	GM	Partly STL
	85	F	
	93	F	Partly ST
	96	F	

April 1944	14	F	
	30	AF	
	37	AF	
	72	AF	
	74	F	
	77A	GM	Partly ST/STL
	85	F	
	93	F	Partly ST

October 1944	14	F	
	30	AF	
	37	AF	
	72	AF	
	74	F	Partly ST
	77A	GM	Partly ST/STL
	85	F	
	93	F	

May 1945	14	F	
	30	AF	
	37	AF	
	72	AF	
	74	F	
	77A	GM	Partly ST/STL
	85	F	
	93	F	

October 1945	14	F	
	30	AF	
	37	AF	
	72	AF	
	74	F	
	85	F	
	93	F	
	137	GM	Until end of month (partly STL)

Appendix Eight:
Analysis of Buses Stored 1940/41 Awaiting Modification

London Terminal Coach Station, 80 Clapham Road, London SW9

(a) Vehicles Taken out of Service from Chelverton Road (31)

Stock No.	Delicensed	Initial Store	Into Main Store	Released	Modified by	Relicensed Date/Garage	
RT 1	1.8.40	F	27.11.40	11.4.41	Chiswick	14.6.41	AF
RT 3	1.8.40	F	27.11.40	11.4.41	Chiswick	23.4.41	AF
RT 5	1.8.40	F	4.12.40	25.3.41	L.T.C.S.	25.3.41	F
RT 6	1.8.40	F	27.11.40	2.4.41	L.T.C.S.	2.4.41	F
RT 16	1.8.40	F	27.11.40	25.3.41	L.T.C.S.	25.3.41	F
RT 17	1.7.40	AF	24.7.40	10.4.41	L.T.C.S.	10.4.41	F
RT 18	1.8.40	F	27.11.40	25.3.41	L.T.C.S.	25.3.41	F
RT 20	1.8.40	AF	4.12.40	15.3.41	L.T.C.S.	15.3.41	F
RT 21	1.8.40	AF	4.12.40	3.2.41*	Chiswick	1.10.41	AF
RT 22	1.7.40	AF	25.7.40	11.4.41	L.T.C.S.	11.4.41	F
RT 25	1.7.40	AF	25.7.40	10.4.41	L.T.C.S.	10.4.41	F
RT 26	1.8.40	AF	27.11.40	10.4.41	L.T.C.S.	10.4.41	F
RT 27	1.7.40	F	25.7.40	9.4.41	L.T.C.S.	9.4.41	F
RT 28	1.8.40	AF	27.11.40	25.3.41	L.T.C.S.	25.3.41	F
RT 29	1.8.40	AF	4.12.40	25.3.41	L.T.C.S.	25.3.41	F
RT 30	1.8.40	AF	27.11.40	11.4.41	L.T.C.S.	11.4.41	F
RT 31	1.8.40	F	4.12.40	11.4.41	Chiswick	18.4.41	F
RT 32	1.8.40	F	27.11.40	4.4.41	L.T.C.S.	5.4.41	F
RT 37	1.7.40	AF	24.7.40	8.4.41	L.T.C.S.	9.4.41	F
RT 38	1.7.40	AF	25.7.40	11.4.41	Chiswick	24.4.41	AF
RT 43	1.8.40	F	27.11.40	25.3.41	L.T.C.S.	25.3.41	F
RT 47	1.8.40	AF	4.12.40	11.4.41	Chiswick	25.4.41	AF
RT 49	1.8.40	AF	4.12.40	4.4.41	L.T.C.S.	5.4.41	F
RT 54	1.7.40	AF	24.7.40	10.4.41	L.T.C.S.	10.4.41	F
RT 56	1.8.40	AF	4.12.40	11.4.41	Chiswick	23.4.41	AF
RT 58	1.7.40	AF	24.7.40	27.3.41	L.T.C.S.	27.3.41	F
RT 61	1.7.40	AF	23.7.40	25.3.41	L.T.C.S.	25.3.41	F
RT 64	1.7.40	AF	25.7.40	10.4.41	L.T.C.S.	10.4.41	F
RT 66	1.7.40	AF	23.7.40	27.3.41	L.T.C.S.	27.3.41	F
RT 67	1.7.40	AF	25.7.40	10.4.41	L.T.C.S.	10.4.41	F
RT 99	1.7.40	AF	24.7.40	5.4.41	L.T.C.S.	5.4.41	F

* RT 21 transferred to Catford Garage on 3.2.41 for use as a guardroom. It was returned to Chiswick on 29.9.41

(b) Vehicles Stored, yet to be Licensed (15)

Stock No	Initial Store	Into Main Store	Released	Modified by	Relicensed Date/Garage	
RT 65	ON	21.12.40	1.4.41	L.T.C.S.	1.4.41	F
RT 71	P	9.10.40	2.4.41	L.T.C.S.	2.4.41	F
RT 75	P	3.10.40	1.4.41	L.T.C.S.	1.4.41	F
RT 80	ON	21.12.40	11.4.41	Chiswick	18.4.41	AF
RT 83	ON	21.12.40	11.4.41	Chiswick	12.4.41	F
RT 101	ON	21.12.40	11.4.41	Chiswick	18.4.41	F
RT 106	ON	21.12.40	11.4.41	Chiswick	12.4.41	F
RT 118	ON	21.12.40	11.4.41	Chiswick	3.5.41	AF
RT 121	ON	21.12.40	11.4.41	L.T.C.S.	11.4.41	F
RT 122	ON	21.12.40	29.3.41	L.T.C.S.	29.3.41	F
RT 124	ON	21.12.40	11.4.41	Chiswick	12.4.41	F
RT 130	P	3.10.40	27.3.41	L.T.C.S.	27.3.41	F
RT 133	ON	21.12.40	11.4.41	Chiswick	23.4.41	AF
RT 140	P	3.10.40	11.4.41	Chiswick	15.4.41	F
RT 141	P	9.10.40	7.4.41	L.T.C.S.	7.4.41	F

Forest Road, Walthamstow, E17

(a) Vehicles Taken out of Service (22) (All Withdrawn on 1.8.40)

Stock No.	Garage	Initial Store	Into Main Store	Released	Modified by	Relicensed Date/Garage	
RT 2	AF	R	28.10.40	11.4.41	Chiswick	6.5.41	AF
RT 12	AF	R	26.10.40	10.9.41	AL	1.10.41	AF
RT 14	AF	RG	10.4.41	12.8.41	AF	1.9.41	AF
RT 15	AF	R	26.10.40	18.8.41	AL	1.9.41	AF
RT 50	AF	ED	24.10.40		Not Recorded	1.9.41	F
RT 52	AF	ED	24.10.40	26.8.41	AL	1.9.41	AF
RT 53	AF	ED	23.8.40	24.8.41	AL	1.9.41	AF
RT 57	AF	RG	10.4.41	18.8.41	AL	1.9.41	AF
RT 70	F	AH	24.10.40	22.8.41	AL	1.9.41	F
RT 90	AF	RG	10.4.41	13.8.41	AF	1.9.41	AF
RT 92	F	ED	23.10.40	24.8.41	AL	1.9.41	F
RT 93	F	AH	21.10.40	23.9.41	AL	1.10.41	F
RT 98	AF	RG	10.12.40	15.8.41	AF	1.9.41	AF
RT 105	AF	RG	10.4.41	20.8.41	AL	1.9.41	AF
RT 116	F	ED	24.10.40	12.8.41	AF	1.9.41	AF
RT 119	F	ED	23.10.40	14.8.41	AF	1.9.41	AF
RT 126	F	ED	24.10.40	3.9.41	AF	12.9.41	AF
RT 127	F	AH	24.10.40	26.8.41	AL	1.9.41	AF
RT 128	F	ED	23.10.40	28.8.41	AL	1.9.41	AF
RT 129	F	HW	16.11.40	21.8.41	AL	1.9.41	F
RT 131	F	HW	22.11.40	15.8.41	AF	1.9.41	AF
RT 135	F	AH	25.10.40		Not Recorded	1.9.41	AF

(b) Vehicles Stored, yet to be Licensed (11)

Stock No	Garage	Into Main Store	Released	Modified by	Relicensed Date/Garage	
RT 76	V,R	30.10.40	21.8.41	AL	1.9.41	F
RT 78	AH	25.10.40	25.8.41	AL	1.9.41	AF
RT 82	AH	24.10.40		Not Recorded	1.9.41	F
RT 91	AH	26.10.40	9.9.41	F	1.10.41	F
RT 104	V,R	26.10.40	24.8.41	AL	1.9.41	F
RT 114	V,R	30.10.40		Not Recorded	1.9.41	AF
RT 123	V,R	30.10.40	28.8.41	AL	1.9.41	AF
RT 132	V,R	30.10.40	26.8.41	AL	1.9.41	AF
RT 135	AH	24.10.40	9.9.41	F	1.10.41	F
RT 138	V,R	26.10.40	17.4.41	Chiswick	3.5.41	AF
RT 144	AH	25.10.40	11.4.41	Chiswick	3.5.41	AF

Windsor L.T. Garage

Vehicles Taken out of Service from Chelverton Road on 1.8.40 (5)

Stock No.	Released	Modified by	Relicensed Date/Garage	
RT 62	16.10.41	AF	1.11.41	GM
RT 86	17.9.41	F	1.10.41	AF
RT 88	17.9.41	F	1.10.41	AF
RT 96	20.9.41	F	1.10.41	AF
RT 97	25.9.41	AH	1.10.41	AF

Staines L.T. Garage

Vehicles Taken out of Service from Chelverton Road on 1.8.40 (4)

Stock No.	Released	Modified by	Relicensed Date/Garage	
RT 107	24.9.41	AH	1.10.41	AF
RT 109	29.9.41	Chiswick	1.10.41	AF
RT 112	29.9.41	Chiswick	1.10.41	AF
RT 115	1.9.41	Chiswick	1.10.41	AF

Reigate L.T. Garage

Vehicles Taken out of Service from Chelverton Road on 1.8.40 (16)

Stock No.	Released	Modified by	Relicensed Date/Garage	
RT 8	11.4.41	Chiswick	11.6.41	AF
RT 33	5.9.41	AF	12.9.41	AF
RT 36	24.2.41*	AF	1.12.41	AF
RT 45	16.9.41	F	1.10.41	AF
RT 63	20.9.41	AH	1.10.41	F
RT 72	20.9.41	AH	1.10.41	F
RT 74	23.9.41	AH	1.10.41	F
RT 81	18.9.41	F	1.10.41	AF
RT 84	19.9.41	F	1.10.41	AF
RT 87	24.2.41*	AH	1.11.41	GM
RT 95	22.9.41	AH	1.10.41	F
RT 113	29.9.41	Chiswick	1.10.41	AF

The following vehicles were initially sent to Reigate (ex-Chelverton Road 1.8.40) but were later sent for further storage to Forest Road, Walthamstow (q.v.):

RT 14 (10.4.41), RT 57 (10.4.41), RT 90 (10.4.41), RT 98 (10.12.40), RT 105 (10.4.41)

* RT 36 was transferred to East Grinstead Garage on 24.2.41 for use as a guardroom. It was returned to Chelverton Road on 10.11.41. RT 87 was transferred to Dartford Garage on 24.2.41 for use as a guardroom. It was returned to Nunhead on 2.10.41.

Potters Bar L.T. Garage

(a) Vehicles Taken Out of Service (11)

Stock No.	Garage	Withdrawn	Into Store	Released	Modified by	Relicensed Date/Garage	
RT 34	AF	1.8.40	1.8.40	10.9.41	F	1.10.41	AF
RT 40	F	1.7.40	27.7.40	11.4.41	Chiswick	3.5.41	AF
RT 41	AF	1.8.40	1.8.40	6.9.41	F	1.10.41	F
RT 42	AF	1.8.40	1.8.40	6.10.41	AH	1.11.41	GM
RT 46	AF	1.8.40	1.8.40	20.9.41	F	1.10.41	AF
RT 51	F	1.7.40	26.7.40	5.9.41	AF	12.9.41	AF
RT 55	F	1.7.40	27.7.40	12.9.41	F	1.10.41	AF
RT 60	F	1.7.40	26.7.40	23.4.41	Chiswick	14.6.41	AF
RT 73	F	1.7.40	27.7.40	16.9.41	F	1.10.41	AF
RT 79	F	1.7.40	27.7.40	15.9.41	F	1.10.41	F
RT 102	F	1.7.40	26.7.40	5.9.41	AF	12.9.41	AF

(b) Vehicles Stored, yet to be Licensed (2)

Stock No.	Garage	Withdrawn	Into Store	Released	Modified by	Relicensed Date/Garage	
RT 77	AF		26.7.40	8.9.41	F	1.10.41	F
RT 111	AF		26.7.40	16.9.41	F	1.10.41	AF

It seems almost unbelievable that the first cine footage in which a 2RT2 bus is visible was shot in colour, its existence being entirely due to a lady by the name of Rosie Newman. Ms Newman was one of the few amateur movie makers who ventured out on to the streets of London to record the aftermath of German bombing raids on the capital and in so doing she not only managed to capture the clearing up operations, but also an unidentified 2RT2 which makes a *fleeting* unscheduled appearance crossing the end of a street. The short sequence is commercially available on the first of the three BBC video cassettes 'The Secret War' towards the end of the programme entitled 'The Battle of the Beams'.

Further 2RT2 wartime appearances on film remain very scarce: however there is an impressive view of a member of the class taken from a second storey window in a documentary film narrated by William Bendix covering the 'friendly invasion' of the United Kingdom by men of the American Eighth and Ninth Airforces.

VE Day 1945 probably brought every available newsreel camera in the country out onto the streets to film the celebrations. Those that were set up at Piccadilly Circus managed to record the occasional glimpse of 2RT2s working on route 14 whose progress was continually impeded by the vast crowds. Many commercial video recordings released to coincide with the 50th Anniversary of VE Day contained such views: a prime example being found in the W H Smith video tape entitled 'D-Day to VE Day' in which a 2RT2, completely surrounded by a throng of joyous Londoners, only advances a few feet.

Numerous newsreel cameras, equipped with both black and white and colour stock, were also used to record images of the Victory Parade which took place in London on 8th June 1946. A recently resurrected colour film of the event includes a view of RT 4 and RT 39 displaying their new liveries and this was transmitted by ITV on the 50th Anniversary of VE Day in 1995.

Production companies engaged in making feature films have been venturing onto the streets beyond the studio front gates since the dawn of Cinema and in so doing have sometimes unconsciously recorded the occasional view of current examples of public transport. However, during the war a considerable amount of filming was accomplished on sound stages and backlots; the footage shot in the capital by newsreel cameras providing a source of material which was used as an inferior alternative to that committed to celluloid on location. There were, of course, notable exceptions such as the 'The Life and Death of Colonel Blimp' (1943, GFD/Archers, colour, directed by Michael Powell and Emeric Pressburger) but nevertheless the chance sighting of buses making unscripted entrances in the background of feature films of the wartime period is regrettably rare.

At the end of this section is a list of films made for general release which contain views of 2RT2s. This is by no means exhaustive, as discoveries continue to be made usually after a rash of rarely-shown films have been televised.

Part of the 'full supporting programme' in the heyday of the cinema was the general interest film which, together with the newsreels, provided additional material between the end of the 'B' movie and the main feature. Amongst these was 'Look at Life' which dedicated one of its programmes to action on the Chiswick skid patch and, as a result, recorded RT 98 in its most famous role. However it was green 2RT2 RT 128 which provided a demonstration on the skid patch for Cliff Richard in a newsreel item which covered his visit to the Training School to master the art of bus driving for his part in the film 'Summer Holiday'.

There are few occasions in which 2RT2s appear in films made or commissioned by London Transport. RT 57 features in two Cine Gazette items, one of which covers the road trials of an RTW test rig seen passing the 2RT2 at Chiswick; the other providing a view of the bus on the move, filmed from another vehicle. And in the film 'Overhaul' the views of post-war RTs arriving and departing Aldenham show an impressive line-up of 2RT2s waiting to act as homeward transport for the majority of the workforce.

Amateurs too were pointing the lenses of their cine cameras at London's buses, probably the most famous group comprising Vic Jones, Geoff Ashwell and Jack Law who succeeded in recording RT 11 in service on route 177 and RT 104 on route 93 on 9.5mm black and white film stock. A considerable amount of their work can be viewed on the Reeltime Films video recording: 'Omnibus, London Archive' which also contains a colour view of an unidentified 2RT2 in revenue service.

Recent feature films set against a background of the Home Front and immediate post-war period have attempted to use authentic forms of transport for the time in which the production was set, with varying degrees of success. The booby prize, however, must go to the director who used a Routemaster in 2RT2 wartime livery in the apparent belief that all London Transport double deck buses are identical.

A specially constructed outdoor set was built for the film 'Hanover Street' (GB 1979, Columbia, directed by Peter Hyams) which had to suffer a simulated enemy bombing raid during which one of the buses would be engulfed in flames. It is interesting to note that the production team had done some homework as RT 44 was used with the two post-war members of the RT family that were altered to represent 2RT2s. To the average cimema-goer the vehicles probably appeared to be correct but to the transport enthusiast, these hybrids with their droopless cabs and red roofs were definitely not the real McCoy. The vehicle that was actually destroyed for the production was, in reality, RTL 1298 which was transformed by Art Department staff who added such cosmetic items as front and rear roofboxes and a central bar and triangle badge to the Leyland radiator. Finally the vehicle was renumbered RT 35 and given the corresponding registration number FXT210 without recognition of the fact that the real RT 35 actually survived the war and met its end at Bird's Commercial Motors of Stratford-upon-Avon in the early sixties. The other 2RT2 stand-in was played by RT 191. A commercial video tape entitled 'London Buses Remembered' (Online Video 1979) contains behind the scenes action of the destruction of 'RT 35'.

'Let Him Have It' (GB 1991, Vivid Films, directed by Peter Medak) also included RT 44 among the collection of members of the RT family sent to Liverpool, which was selected to represent Croydon in the early '50s. Unfortunately, RT 44's wartime colour scheme appeared totally out of place for the time in which the film was set and one wonders whether a member of the production team should also have suffered death by hanging for failing to do some essential research into London bus liveries.

Feature Film Appearances by 2RT2s

Love Story, GB, 1944, Gainsborough, black and white. Leslie Arliss

I Live in Grosvenor Square, GB, 1945, ABP, black and white. Herbert Wilcox

Wanted for Murder, GB, 1946, Marcel Hellman, black and white. Lawrence Huntington

Night and the City, GB, 1950, 20th Century Fox, black and white. Jules Dassin

Seven Days to Noon, GB, 1950, London Films, black and white, John Boulting

Laughter in Paradise, GB, 1951, Transocean, black and white. Mario Zampi

Double Exposure, GB, 1954, Rank/Kenilworth, colour. John Gilling

Knock on Wood, US, 1954, Paramount, colour. Norman Panama and Melvin Frank

The Constant Husband, GB, 1955, British Lion/London Films, colour. Sidney Gilliat

The Ladykillers, GB, 1955, Ealing, colour. Alexander Mackendrick

23 Paces to Baker Street, US, 1956, 20th Century Fox, colour. Henry Hathaway

Hidden Homicide, GB, 1959, Bill and Michael Luckwell, black and white. Tony Young

Although 2RT2s were withdrawn from Central Area passenger service towards the end of May 1955, their appearances in feature films after that date are entirely due to stock shots or previously filmed footage being used. 'Knock on Wood' contains one of the biggest continuity errors of all time wherein Danny Kaye, the star of the film, hides in the back of a van parked in a narrow lane leading off Charing Cross Road. The van then drives away and arrives, ostensibly from the same narrow lane, in Fleet Street where, in the background, a 2RT2 can be seen heading towards Ludgate Circus on route 96.

Appendix Ten:
Total Numbers of Vehicles in Stock, 1939-1964

Month/Year	Delivered	Withdrawn	Relicensed	Total in Stock	Total Licensed (PSV)
8/39	1(a)			1	1
1/40	15			16	16
2/40		1(b)		15	15
3/40	40			55	55
4/40	5			60	60
5/40	31			91	91
6/40	17			108	108
7/40		20(s)		108	88
8/40		68(s)		108	20
9/40	2		3	110	25
12/40				110	25
3/41	5		10	115	40
4/41	12		19	127	71
5/41	3		2	130	76
6/41	1		3	131	80
9/41	7		22	138	109
10/41	9		25	147	143
11/41			3	147	146
12/41	2		1	149	146
2/42	1			150	150
10/42			1(b)	151	151
12/42				151	148
12/43				151	148
12/44				151	146
8/45		1(a)		150	145
12/45				150	149
12/46				150	146
1/47		1(c)		149	146
12/47				149	140
12/48				149	128
3/49	1(d)			150	128
6/49		1(e)		149	127
12/49				149	120
12/50				149	124
6/51		1(f)		148	122
12/51				148	121
12/52				148	131
12/53				148	138
4/54		1(b)		147	129
12/54				147	128
3/55		1(d)		146	60
6/55		1(g)		145	7
12/55		18(h)		127	7
12/56				127	7

Month/Year	Withdrawn	Total in Stock
12/57		127
12/58		127
8/59	1	126
12/59		126
2/60	2	124
3/60	4	120
4/60	6	114
6/60	1	113
9/60	40	73
10/60	1	72
12/60	1	71
1/61	1	70
2/61	1	69
12/61		69
8/62	1	68
11/62	6	62
12/62	1	61
2/63	2	59
3/63	4	55
4/63	12	43
5/63	7	36
6/63	11	25
7/63	6	19
8/63	7	12
9/63	9	3
12/63	1	2
3/64	2	0

(a)	RT 1
(b)	RT 19
(c)	RT 97
(d)	RTC 1
(e)	RT 85
(f)	RT 19
(g)	RT 59
(h)	Includes RT 106
(s)	Stored